T0315050

ELECTRICITY MARKETS

ELECTRICITY MARKETS
Theories and Applications

Jeremy Lin
Fernando H. Magnago

IEEE
PRESS
SERIES
ON POWER
ENGINEERING

IEEE PRESS

WILEY

Published by John Wiley & Sons, Inc., Hoboken, New Jersey.
Published simultaneously in Canada.

No part of this publication may be reproduced, stored in a retrieval system, or transmitted in any form or by any means, electronic, mechanical, photocopying, recording, scanning, or otherwise, except as permitted under Section 107 or 108 of the 1976 United States Copyright Act, without either the prior written permission of the Publisher, or authorization through payment of the appropriate per-copy fee to the Copyright Clearance Center, Inc., 222 Rosewood Drive, Danvers, MA 01923, (978) 750-8400, fax (978) 750-4470, or on the web at www.copyright.com. Requests to the Publisher for permission should be addressed to the Permissions Department, John Wiley & Sons, Inc., 111 River Street, Hoboken, NJ 07030, (201) 748-6011, fax (201) 748-6008, or online at http://www.wiley.com/go/permission.

Limit of Liability/Disclaimer of Warranty: While the publisher and author have used their best efforts in preparing this book, they make no representations or warranties with respect to the accuracy or completeness of the contents of this book and specifically disclaim any implied warranties of merchantability or fitness for a particular purpose. No warranty may be created or extended by sales representatives or written sales materials. The advice and strategies contained herein may not be suitable for your situation. You should consult with a professional where appropriate. Neither the publisher nor author shall be liable for any loss of profit or any other commercial damages, including but not limited to special, incidental, consequential, or other damages.

For general information on our other products and services or for technical support, please contact our Customer Care Department within the United States at (800) 762-2974, outside the United States at (317) 572-3993 or fax (317) 572-4002.

Wiley also publishes its books in a variety of electronic formats. Some content that appears in print may not be available in electronic formats. For more information about Wiley products, visit our web site at www.wiley.com.

Library of Congress Cataloging-in-Publication Data is available.

ISBN: 978-1-119-17935-1

Printed in the United States of America.

10 9 8 7 6 5 4 3 2 1

To my parents
[Jeremy Lin]

and

To Gaby, Pris, and Andy
[Fernando H. Magnago]

Contents

Contents

About the Authors

Dr. Jeremy Lin has more than 18 years of experience in power system planning, operations, and markets. He has extensive knowledge about industry restructuring and electricity market developments in the United States. He also has significant experience in modeling, simulation, analysis of restructured electricity market, transmission system analysis, power flow analysis, and advanced computer technology applications to power system. He is currently affiliated with PJM Interconnection. Dr. Lin received his MSEE in power and energy systems from University of Illinois at Urbana–Champaign and Ph.D. in electric power engineering from Drexel University. He has published numerous publications in both top-ranked journals and conference proceedings. Currently, he has various collaborative research works with many researchers from domestic and international institutions. He is a senior member of IEEE.

Dr. Fernando H. Magnago has more than 28 years of experience in research and development in power system software and devices. He is a professor in the Department of Electrical and Electronic Engineering at National University of Rio Cuarto, Argentina, working on undergraduate, graduate, and research projects related to power system analysis, optimization, and production simulation. He has designed and developed analytical techniques and softwares principally related to fault analysis, state estimation, and security-constrained unit commitment. Dr. Magnago is the author of 3 books, a book chapter, 22 journal papers, and 70 conference publications in the area of power systems.

Preface

Why did an electricity market emerge? How does it really work? What are the performance measures that we can use to tell that the electricity market under consideration is well functioning? These are the questions that will be explored in this book. The main purpose of this book is to introduce the fundamental theories and concepts that underpin the electricity markets which are based on three major disciplines: electrical power engineering, economics, and optimization methods.

This book is intended for first-year graduate students or senior-year undergraduate students as well as practitioners and professionals who would like to study the fundamental and advanced topics in electricity markets. The backgrounds of students or readers should not necessarily be restricted to the disciplines mentioned above. Emergence of markets in many engineering disciplines, such as electrical power engineering and telecommunication engineering, challenges us to draw upon the theories and concepts beyond the realm of such engineering disciplines. A good understanding of these fundamental concepts is necessary to further enhance the understanding of the complex operation of power system and electricity markets.

Representation of physical systems by mathematical equations is fairly common in many engineering disciplines. This book is no exception to this tradition. However, the authors try to have a level of mathematical sophistication so that the material is accessible to many students and practitioners as well as the important issues will be treated in a rather meaningful way.

The first chapter of the book will briefly describe the nature and characteristics of electric power system, and cover the basic drivers for the transformation of electricity industry in the United States and around the world. With that tremendous change brings challenges and complex issues. Understanding these complex issues requires both basic and advanced knowledge of electrical power engineering principles, microeconomic theories, and basic and advanced optimization methods. Therefore, fundamentals of electric power system are treated in Chapter 2, and relevant microeconomic theories are introduced in Chapter 3.

The key components of power system economic operation—unit commitment, economic dispatch, and optimal power flow—are covered in Chapters 4, 5, and

6. After having a solid understanding in the areas of how the power system is operated, students will be exposed to the fundamental elements of electricity markets to understand how electricity markets are designed and structured. For this reason, we will cover the design, structure, and operation of an electricity market in Chapter 7.

So, what is the outcome, intended or unintended, of the electricity market operation explained in the previous chapter? The topics that will be covered in Chapter 8 include market pricing, such as zonal pricing versus nodal pricing, market modeling, and simulation of electricity markets which have wide applications in the industry. With the emergence of any electricity market, the issue of market power and its mitigation, market performance, and other issues naturally appear. These issues are the inevitable results of a deliberate coalescence of economic theories and physical law-based electric power system. The fundamental approaches and methods used to evaluate an electricity market, particularly detecting and mitigating market power are broadly treated in Chapter 9. Students who have basic understanding of these topics will be in a better position to understand more complex issues that subsequently arise from the operation of electricity markets.

In Chapter 10, we will deal with one of the critical issues, that is, the system planning under the context of electricity market regime. We will provide reasons why there are new ways to solve the same problem, because the electric power system which has operating electricity markets has tremendous amount of economic data related to power system at its disposal. These new data will enable us to solve the same problem in new ways. Another emerging issue is the role that the electricity market will play under smart grid and microgrid environments. We will also provide some qualitative arguments related to those emerging, yet important, topics in the final chapter, Chapter 11.

While the book covers several topics related to electricity markets and the fundamental theories behind them, it is a challenge for an instructor to cover all the chapters in this book for a one-semester course. It is therefore necessary to make some choices as to cover which chapters that are deemed important for the students. The authors would also like to acknowledge that it is not possible to include all relevant references and sources for all the materials covered in this book. However, interested readers can explore more by tracking additional references which are outside of these given in the further reading section of each chapter.

The authors would like to express their sincere gratitude towards many colleagues at various organizations, academic and research institutions, and their former academic advisors. The authors particularly would like to thank Herminio Pinto from Nexant, and Diego Moitre and Juan Alemany from GASEP, Argentina. This book could not have been complete without previous discussions and research work done with them. The authors would also like to thank reviewers who made valuable comments and suggestions in the first manuscript which helped improve the final content of the book.

The book is dedicated to anyone who is and will be fascinated by the complex operation of electricity markets both in the United States and around the world.

JEREMY LIN
FERNANDO HUGO MAGNAGO

Chapter 1

Introduction

This beginning chapter will provide a high-level overview of the topics related to the basic drivers and transformation of electricity industry around the world. With that tremendous change come challenges and complex issues. One of the key developments in this transformation is the development of electricity markets. This is the main topic of this book. In fact, the primary and paramount goal of an electric power system operation is to maintain a high level of reliability. Under a restructured environment, this goal of system reliability is achieved via a market mechanism. Understanding electricity markets requires both basic understanding and advanced knowledge of electrical power engineering principles, microeconomic theories, and optimization methods from the field of operations research. Therefore, the fundamentals of these topics will be covered in the first few chapters.

1.1 ELECTRIC POWER SYSTEM

Electricity is indispensable for a modern society. The marvels of a modern life that we enjoy today cannot be possible without electricity. The importance of electricity is without questions. So, how do we get electricity?

In general, electricity is generated from electric generating sources located far from the load centers, then transmitted over long distances using transmission lines, and distributed to the load customers which include factories, offices, and homes. The entire chain of this system is known as *electric power system*.

The electric power system as we know of today was developed more than a 100 years ago. It generally consists of generation, transmission, and distribution subsystems. The entire chain of business from generation to transmission and distribution to load customers for a particular service area is owned by a single entity, known as an *electric power company* or *electric utility company*. The electric power company

Electricity Markets: Theories and Applications, First Edition. Jeremy Lin and Fernando H. Magnago.
© 2017 by The Institute of Electrical and Electronics Engineers, Inc. Published 2017 by John Wiley & Sons, Inc.

is either owned and operated by a national government or can be a public company owned by investors, but operated by management and employees of the company. Therefore, electric power industry is an important part of a country's infrastructure.

For the last two decades, the electricity industry around the world has undergone a tremendous transformation. This transformation is from a traditional structure of an electric power industry typically owned by national governments or public investors (as in investor-owned utilities) towards a structure that is exposed to a competitive market environment. This transformation in electricity industry was preceded by similar transformations in other industries such as airlines, trucking, and natural gas industries. In the case of an electric utility owned by a national government, this transformation is in the form of privatization first, then the privatized company is exposed to an open, competitive market environment. This was the case for electricity industry restructuring in Argentina. In the case of an investor-owned utility as in the United States, the generation part of the business was separated from the wires part (transmission and distribution systems) of the same company. This generation business is either divested to an independent company or completely formed as a separate subsidiary of the original utility. This is equivalent to the functional unbundling of an existing vertically integrated company.

The key outcome of this entire industry transformation, which is generally known as *electricity industry restructuring*, is the development and establishment of electricity markets. This is achieved by breaking up generation services into a separate, more competitive segment of the industry while the transmission and distribution parts of the utility service largely remain a regulated monopoly service. Because of unbundling of services (generation vs. transmission and distribution), these services have to be priced separately on a customer's bill.

Generally, in an electricity market, generators (generator owners) compete among each other to have an opportunity to supply electricity to serve load customers at the other end of the wires. However, the transmission and distribution parts of the system (electric power system) are not open for competition because it is generally believed that the wires business is subject to a natural monopoly behavior. A firm with natural monopoly enjoys significant economies of scale. It has to be properly regulated due to potential market power issues.

Therefore, the competition among generators is one of the key developments when the electric power industry was restructured. As a consequence, the analysis of the strategic interactions among the competing generators becomes an important subject to explore. These topics will be covered in more detail in later chapters. Natural monopoly part of the system, that is, transmission and distribution system, is still regulated because it will create more inefficiencies in the system if more than one firm are allowed to compete for wires business. One way to effectively regulate the network system is to form an independent entity that would control and operate the network only with or without the ownership of these facilities. While such entities may carry different names, such as independent system operator (ISO), transmission system operator (TSO), or regional transmission organization (RTO), the key functions of these entities are essentially the same. The mandate of these entities is to operate and manage the network system in a fair, least-cost, and most-efficient manner so that generators can compete effectively on a level playing field. The primary goal of operating the network in such a manner is to increase the economic efficiency of the system, and thus increase the social welfare. However, there might be some variations among these system operators in the areas of ownership, non-profit or for-profit, financial and capital structures, and governance. The list here is not exhaustive, but just descriptive. These topics are beyond the scope of this book.

In the electricity market setting, electricity is treated as another commodity. However, electricity must be generated simultaneously with demand which constantly fluctuates. As a result, an additional capacity, called the reserve margin, must be available to compensate for planned and unplanned outages of generating plants as well as spikes in demand. In a sense, the unique characteristics of electricity provide challenges unlike any of the other industries that has been deregulated. Sometimes, some sort of intervention is needed to ensure adequate supply. There are some imperfections in the competitive wholesale market operations, so some kind of reforms or interventions are generally needed.

1.2 ELECTRICITY INDUSTRY RESTRUCTURING IN THE UNITED STATES

The electric power system in North America is divided into three large regions: *Eastern Interconnection*, *Western Interconnection*, and *ERCOT Interconnection*, where ERCOT stands for Electric Reliability Council of Texas, as shown in Figure 1.1. Each interconnection operates its own system with small ties to other interconnections. The nominal system frequency for the entire system, including all three interconnections, is 60 Hz.

1.2.1 Key Drivers for Electricity Industry Restructuring

It is generally believed that the following are the key reasons and drivers behind the electricity industry restructuring:

1. Technological changes
2. High electricity costs
3. Overall system inefficiencies
4. Higher environmental restrictions

Technological changes have been an important driver to allow the implementation of competitive schemes in an industry which had been historically considered as a natural monopoly. New technologies make it economical for competitors to provide

North American Electric Reliability Corporation Interconnections

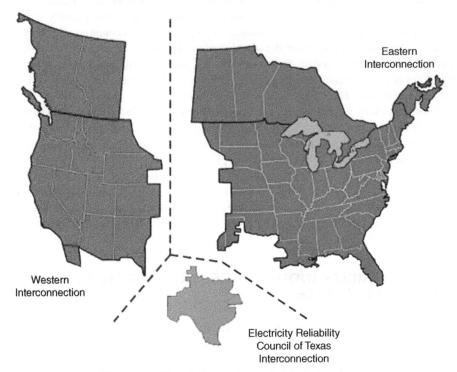

Figure 1.1 Electric Power System in North America.

electric generation services to electricity consumers, particularly industrial customers, which were traditionally served by incumbent utilities. Legal authority given to large industrial users to bypass the local utility provided more impetus to the industry restructuring.

In large part, the action was prompted by the burden of having a higher or the highest electricity costs in the country, which created hardships for residential consumers and handicapped many businesses from competing on a "level playing field" with companies located outside the region. For example, New England was one of the first regions of the country to restructure the industry. The persistently high cost of electricity which put the region at a competitive disadvantage was another driving force behind the push for further competition in the generation sector.

Another key driver in restructuring efforts was the environmental protection whose goal is to reduce atmospheric emissions from generating electricity. More rigorous air emission standards and regulations led to the construction of new natural gas-fired generating plants which, in turn, led to emission reduction and air-quality improvements. The new cleaner generators have displaced the older, inefficient, and polluting generating plants. Environmental protection rules, such as the Clean Air Act and subsequent federal rules along with the state's air quality regulations led to increased environmental benefits. The key pollutants that caused global environmental issues are sulfur dioxide (SO_2) which is responsible for acid rain, nitrogen oxide (NO_x) which produces smog, and carbon dioxide (CO_2) which is one of the key drivers of global warming.

The main objective of industry restructuring was to create a fair and reliable market for competition in generating electricity while ensuring equal access to transmission grids. The other objectives were to achieve lower electricity rates and enhance economic growth. Once established, the wholesale market treats electricity as a commodity with prices set not by regulators, but by market rules and the balance between supply and demand.

1.2.2 Pre-Federal Energy Regulatory Commission Order 2000

Traditionally, the majority of the electric utilities in the United States is formed as investor-owned utilities (IOUs). Some utilities are owned by the federal government and some are owned by municipalities and cooperatives. The vertically integrated utilities are granted franchise areas with the exclusive right to provide electric service. In exchange for this monopoly right, almost every aspect of their business was regulated by state's public utility commissions (PUCs) within their state boundaries. State PUCs set the operating standards for electric service, authorize the utilities to invest in new facilities such as power plants, transmission lines, or other equipments needed to meet their customer service obligations, and set the rates that customers pay for electricity service. Electric utilities are responsible for supplying electricity to load customers in their service territories.

The US Congress laid the groundwork for deregulating or restructuring wholesale electricity markets through provisions contained in the Public Utility Regulatory Policies Act of 1978 (PURPA). The Act mandated that regulated electric utilities provide a market for the output of non-utility generating (or power) plants that meet certain size, technology, and environmental criteria. Many state regulators required utilities to sign long-term power purchase contracts with small, independent PURPA generators at the utilities' then-avoided costs. Power plants that were built pursuant to PURPA represented the beginning of a new class of generators called independent power producers (IPP). Furthermore, pursuant to the state-mandated integrated resource planning processes, regulators required utilities to compare the cost of utility-built generation with that from IPP's generation and to take the least-cost alternative. This regulatory paradigm resulted in the maturation of the IPP industry across the country.

Thereafter, Congress passed the Energy Policy Act of 1992 which advanced the move to competition in wholesale markets. The Act gave Federal Energy Regulatory Commission (FERC) an authority to order utilities to provide transmission access to third parties in the wholesale electricity markets. This began the process of allowing open access to the existing transmission system to non-utility generators. This also created a condition in which there were increased competitions among generators owned by electric utilities and IPPs.

Subsequently, FERC issued Orders 888 and 889 in April 1996, which authorized open and equal access to all utilities' transmission lines for all electricity producers, thus facilitating wholesale and retail restructuring. These orders called for an accurate calculation and posting of available transfer capability (ATC) and implementation of the Open Access Same-Time Information System (OASIS), requiring transmission owners to open their transmission systems to third parties, giving equal access and fairness to use their transmission facilities to transfer power. In addition to asserting federal jurisdiction over all transmission, FERC Order 888 states that transmission-owning utilities must charge competing utilities the same fees to use their transmission system as they charge themselves. For the most part, FERC sets the transmission rates for wholesale transactions among transacting parties.

A few independent system operators (ISOs) were established to help achieve these and other objectives, although most transmission systems continued to be operated by the owning utilities. The wholesale portion of the US electric power industry has been shaped by FERC major orders dealing with electricity transmission: FERC Order 888–889, and subsequently Order 2000.

1.2.3 Post-Federal Energy Regulatory Commission Order 2000

Despite the push by FERC orders towards more open and fair access to the nation's transmission system, there were evidences that suggested that there were continued discrimination in the provision of transmission services by vertically integrated utilities against other users of their transmission system. This result may be impeding fully competitive electricity markets. That also implies that these orders failed to fully achieve what they were supposed to accomplish, that is, increased competitiveness of open electricity markets.

Frustrated by these outcomes, which are impediments to open competition, FERC later issued far-reaching "Order 2000" in December 1999 to push and expedite the development of efficient electricity markets by further calling for the formation of regional transmission organizations (RTOs) in various parts of the country. FERC Order 2000 requires FERC jurisdictional utilities to either file a plan by October 15, 2000, to establish an RTO whose function is to independently operate the transmission systems, or if a filing is not made, then each utility must explain why they are not making such a filing.

As envisioned by FERC, RTOs will implement and operate efficient electricity markets and also manage and operate the nation's transmission grid. FERC believes that RTOs, if established, will bring about the following benefits: increased efficiency, improved congestion management, accurate estimates of total transfer capability (TTC) and ATC, efficient planning of transmission and generation, increased coordination among states, and reduced transaction costs. All of these benefits will help promote competition and efficiency in wholesale electricity markets. The major role of RTOs is to provide fair and reasonable access to the transmission network nationwide. In consequence, electricity consumers would be expected to pay the lowest price possible for reliable service. FERC Order 2000 is a defining moment in the history of electric power system in the United States.

1.2.4 Regional Transmission Organization

RTOs, as called for by FERC, must have four minimum characteristics:

1. Independence
2. Scope and regional configuration
3. Operational authority
4. Short-term reliability

RTOs must also perform eight minimum functions:

1. Tariff administration and design
2. Congestion management
3. Parallel path flow
4. Ancillary services
5. OASIS and TTC/ATC
6. Market monitoring
7. Planning and expansion
8. Interregional coordination

Under Order 2000, the formation of RTOs is voluntary, and their organizational form is quite flexible. RTOs can take the form as not-for-profit ISOs or for-profit TransCo models. The focus is on "characteristics" and "functions." One of the salient characteristics is that the RTO must serve an appropriate region that must be of sufficient scope and regional configuration to permit the RTO to maintain reliability and effectively perform its required functions.

In general, RTO is a voluntarily formed entity that ensures comparable and non-discriminatory access by electric generators to regional electric transmission systems.

RTOs are governed in a manner that renders them independent of the commercial interests of power suppliers who may also own transmission facilities in the region. The RTO assumes operational control of the use of transmission facilities, administers a system-wide transmission tariff applicable to all market participants, and maintains short-term system reliability.

Based on these characteristics, FERC proposed three RTOs in the Eastern Interconnect region: one in the Midwest, one in the Northeast, and one in the Southeast. In the Northeast, FERC tried to facilitate the process of merging three existing ISOs: PJM, New York ISO (NYISO), and ISO New England (ISO-NE), although these efforts were ultimately terminated without achieving an integrated Northeastern RTO. In terms of "scope and regional configuration" characteristics, FERC has not defined geographical boundaries for RTOs, leaving it to the transmission owners to determine appropriate consolidations that are sufficiently regional in size and scope. To date, the FERC-approved RTOs include Midcontinent ISO (MISO), PJM, Southwest Power Pool (SPP), California ISO (CAISO), NYISO, and ISO-NE in the United States. The transmission grid that the ERCOT ISO administers is located solely within the state of Texas and is not synchronously interconnected to the rest of the United States. The transmission of electric energy occurring wholly within ERCOT is not subject to the Commission's jurisdiction. The regional boundary map of RTOs/ISOs in the United States and Canada as of November 2015, is shown in Figure 1.2.

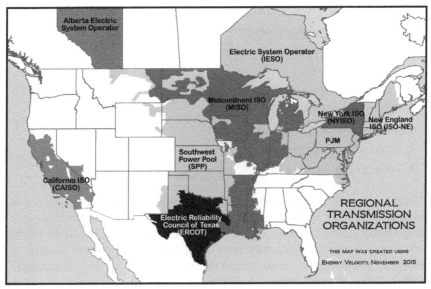

Figure 1.2 Map of Regional Transmission Organizations in North America. *Source:* http://www.ferc.gov/industries/electric/indus-act/rto.asp. Public domain.

1.2.5 Post-Regional Transmission Organization

In July 2002, FERC issued its Notice of Proposed Rulemaking on Standard Market Design (SMD NOPR), endorsing a market design that incorporates many of the best practices of the PJM and NYISO markets, such as *locational marginal pricing* (LMP) and *congestion revenue rights* (CRRs). An LMP scheme is currently used in PJM, NYISO, ISO-NE, and MISO markets. CRRs are also known as *financial transmission rights* (FTRs) in PJM and *transmission congestion contracts* (TCCs) in NYISO. The SMD policy includes far more implications for regional electricity markets than the previous RTO policy (Order 2000) on issues such as power rates, resource planning, and demand management.

FERC issued its NOPR on SMD to establish a uniform market structure and rules for emerging electricity markets. Any market participant who has to buy or sell electricity across two RTOs must currently follow two different tariffs, rules, and protocols. Inconsistent rules across the United States cause market inefficiency and raise costs for customers. FERC believes that SMD, if adopted, will create consistent market rules administered by fair and independent entities, no matter where the electricity is bought or sold. As a consequence, FERC expects SMD to lower costs to customers, to eliminate residual discrimination, to protect against potential market manipulation through market power mitigation measures and oversight, and to create incentives for investment in electric infrastructure, that is, transmission, generation, and demand-side resources with clear transmission policy and planning policies for grid expansion. The major requirements of FERC-proposed SMD are

1. Independent transmission provider (ITP)
2. Independent transmission companies (ITC)
3. Single transmission tariff
4. Long-term bilateral contract market
5. Day-ahead and real-time markets for energy and ancillary services
6. Regional transmission planning
7. LMP for congestion management
8. CRRs for tradable transmission rights
9. Market power monitoring and mitigation
10. Regional resource adequacy requirement
11. Role of states
12. Governance

However, on July 19, 2005, FERC issued an order terminating the SMD proceeding. This was partly due to the numerous criticisms about overreaching its federal authority over various issues such as power rates, demand forecasting, resource planning, and demand-side management.

In the next sections, electricity industry restructuring in some countries in Latin America, Europe, and Asia are presented.

1.3 ELECTRICITY INDUSTRY RESTRUCTURING IN LATIN AMERICA

Electricity industries in Latin America have undergone an enormous transformation in the 1990s. Basically, three electricity markets were developed in the region: the Central American market, the Andean market, and the Common Market of the Southern Cone. The Southern Cone market is the largest market including Chile, Argentina, and Brazil. The electricity industry restructuring in these markets are briefly described below. The electricity market in Mexico was only recently developed.

1.3.1 Chile

Chile is known as the pioneer of electricity industry reform. Before the restructuring in 1980s, the electric power industry in Chile has a vertically integrated structure which is centrally planned and heavily regulated. The centralized planning and operation were replaced by market-oriented approaches by deregulation and privatization. The objective of electric restructuring is to establish conditions for economic efficiency and to attract private investments.

Chilean Ministry of Energy has jurisdictional authority over the electricity sector. It is responsible for plans, policies, and standards regarding the development of the energy sector. In addition, it grants concessions for hydroelectric power plants, transmission lines, and distribution areas. Under the Ministry of Energy, the *Comisión Nacional de Energía* (CNE) is a technical agency responsible for studying prices, tariffs, technical standards, and setting tariffs according to the applicable regulations, and generating the electrical infrastructure work plan.

As part of the electricity industry restructuring in Chile, a wholesale electricity market in which the generation sector was opened for competition was developed. The business of transmission and distribution is still regulated due to their monopolistic characteristics. The transmission system is open for access by any legitimate market participants. Network costs are socially allocated. The distribution system is regulated based on some incentives.

In this market environment, at least large electricity consumers are exposed to unregulated market prices while the smaller consumers are protected from price volatility with some pass through of wholesale market prices. The so-called *Poolco* model in which generators compete under centralized generation dispatch was introduced. The market utilizes two-part pricing schemes: energy pricing and capacity pricing. In energy pricing, short-term marginal cost of energy is used as part of nodal pricing which considers both generation and transmission constraints. For capacity pricing, capacity payments are made to generators which make their capacities available in the yearly peak demand period (from May to September). The capacity payment depends on availability, start-time, and time to reach a full load energy production. The capacity price is defined by the regulator every 6 months based on the fixed cost of a typical gas-turbine generator.

In the energy market, financial (non-physical) bilateral contracts are also allowed. This is equivalent to the virtual bids which are eligible to participate in the electricity markets in the United States. The generation is centrally dispatched by *Centro de Despacho Económico de Carga* (CDEC). The economic dispatch is based on hourly marginal cost which is based on the variable costs of thermal generators. The variable cost of thermal generators are audited and hydro units are dispatched based on the cost of water estimated by CDEC. Economic transactions among generators are done based on marginal costs. It has entered into a second stage of reforms with public power purchase agreement (PPA) auctions in a private environment.

1.3.2 Argentina

Argentina began to reform its energy sector as part of a wider economic reform in the early 1990s. The long history of inefficient performance of state-owned utilities was one of the key drivers for the major transformation in the energy sector. The "Electricity and Natural Gas Acts" passed in the early 1990s paved the way for a new regulatory framework. Consequently, state-owned utilities were unbundled both horizontally and vertically as well as privatized. Wholesale markets for both electricity and natural gas were developed. The electricity industry is regulated by an authoritative regulatory body, *Ente Nacional Regulador de la Electricdad* (ENRE).

Private companies are allowed to participate in electricity generation which was open for competition while transmission and distribution parts of the system were still treated as regulated monopolies. The primary objectives of electricity restructuring in Argentina were to reduce electricity tariff, improve the quality of service, expand consumers' choices, and improve the economic efficiency. Due to the significant investments in the generation sector and some investments in transmission in the late 1990s, the Argentinean electricity market was one of the most competitive markets globally. The Argentinean model becomes a benchmark for measuring the success of electricity restructuring throughout the world.

To facilitate the development of a wholesale electricity market, an independent system operator (ISO), known as *Compañía Administradora del Mercado Mayorista Eléctrico S.A.* (CAMMESA) was established. CAMMESA is both a market operator and a market administrator. It provides open access to the market and transmission system to every market participant and establishes market rules. Its main roles entail delivery coordination, responsibility for wholesale price setting, and management of the economic transactions done by *Sistema Argentino De Interconexión* (SADI) which is the main transmission system.

The Argentinean electricity market comprises of both forward (contract) market and spot market. In the forward market, generators and distributors or large users can negotiate freely and sign contracts for electricity which set both prices and quantities for future delivery. In real-time, the electricity users with firm contracts are given priority in the event of shortages provided that the contracted generator is available. In the spot market, energy prices are set hourly on each system node based on the short-run marginal cost. Nodal energy prices also reflect marginal losses produced by generation/load and transmission congestion with local (zonal) pricing. The market does not have transmission rights, such as FTR. Ancillary services are both regulated and market based. For example, generators are obligated to provide frequency regulation (primary and secondary). But, they can trade their obligations among themselves. To provide voltage support, each market participant has to have a sufficient level of reactive power. Deviations from standard operating levels are penalized. Payments for black start services are based on regulated price. Transmission companies are not allowed to trade energy.

1.3.3 Brazil

Restructuring of electric power industry in Brazil followed a similar pattern as those in other Latin American countries. Before the restructuring, electric utilities are owned by the national government, which guarantees a certain level of rate of return. This leads to overinvestment and inefficiencies in the system. Electric utilities own all sectors of the business: generation, transmission, distribution, and retailing. Financial crisis in 1999 led to the payment default of sectoral liabilities and shortage in investment. Brazil started its power sector reform in 1996. The key objectives of this reform are to (1) ensure supply through continuity of expansion, (2) maintain and improve efficiency, (3) provide better service and competitive price setting, (4) provide more choices for consumers, and (5) reduce government debt through nongeneration of new debts and asset privatization.

The new rules, set forth in the restructured environment, were designed to introduce competition in generation and retailing sectors. However, the wires part (transmission and distribution) was still regulated with provisions for open access due to its monopolistic nature. As the result of restructuring, a wholesale energy market was developed, along with an establishment of an independent system operator, *Operador Nacional do Sistema Elétrico* (ONS), to facilitate the competition. A regulatory agency, *Agência Nacional de Energia Elétrica* (ANEEL), was also established and most distribution utilities were privatized. The wires part is subject to revenue cap and yardstick competition. For generators and retailers which are exposed to competition, the return on investment is based on their ability to manage risk under stable market rules.

Under its wholesale energy market regime, both generation and transmission resources are centrally dispatched on the least-cost basis by the system operator. There were no market rules or mechanisms that allow generators and load to offer and bid into the market based on price. Hydro units are dispatched based on their expected opportunity costs which are computed by a multistage stochastic optimization which models the detailed representation of hydro plant operation and inflow uncertainties. Bilateral contracts or other commercial arrangements are not considered in the centralized dispatch.

Market-clearing prices in the wholesale energy market are represented by short-run marginal costs calculated from the Lagrangian multipliers of the stochastic dispatch model. With any electricity market based on short-run marginal cost, the missing money problem is inevitable. In the missing money problem, a certain set of generators did not receive sufficient amount of revenues from the energy market to sustain their business viability. The revenue insufficiency for generators created by short-term spot price leads to an inability to provide sufficient incentives for new generation. In the Brazilian energy market, the prices are generally either volatile or very low due to the predominantly hydro system. The system marginal costs (spot prices) become low when there are surplus energy in the system, and become high when there is a very dry period or drought.

To encourage healthy entrance of new generation, a scheme based on mandatory bilateral contracts was introduced. First, all loads are required to be fully covered by power purchase agreements (PPAs). Second, these financial forward contracts must also be firmed up by actual generation capacity similar to firm energy from hydro plants. Therefore, such new contracts are used as a mandatory mechanism to secure energy supply for potentially growing load so as to facilitate the entry of new generators. To improve the long-term efficiency of the industry, those PPAs are also arranged through competitive auctions.

1.3.4 Mexico

The planned restructuring activity currently underway in Mexico is a prime example to show that more and more countries are interested in opening up their power industries to enjoy the economic benefits that can be potentially brought about by industry restructuring which encourages competition.

The new Law of the Electric Industry (*Ley de la Industria Eléctrica*), effective August 12, 2014, would allow the private sector to participate freely in the generation and sale of electricity while the electric grid will still be under the operational control of a state-owned agency. The new electric industry law will create a new wholesale electricity market (*Mercado Eléctrico Mayorista*) to be operated by the national energy control center, *Centro Nacional de Control de Energía* (CENACE), currently a unit within the federal commission, *Comisión Federal de Electricidad* (CFE). CENACE will also become the independent system operator for the entire grid. Mexican Ministry of Energy, *Secretaría de Energía de México* (SENER), and the regulatory authority, *Comisión Reguladora de Energía* (CRE) will have regulatory oversight and supervisory authority over the wholesale power market.

Under the new law, SENER crafted a draft regulation document on the guidelines for the electricity market, *Bases del Mercado Eléctrico*, and sent it to the Federal Commission for Regulatory Improvement (COFEMER) in February 2015. COFEMER is required to conduct a cost/benefit analysis of this new regulation. The guidelines establish the principles for the design and operation of the wholesale electricity market (WEM) including auction rules. All regulations before COFEMER are subject to public comment. After the review and decision by COFEMER, SENER will issue the final guidelines which will become a detailed plan for developing and operating an electricity market.

The key topics contained in the guidelines include staged implementation of the market, system reliability, market operation, operational planning, long-term markets, market monitoring, credit, and billings. The final guidelines will turn into the major protocols for the wholesale market operation. These protocols include (1) "Market Practice Manual" which will describe the principles for instructions and procedures for the administration, operation, and planning of the WEM, (2) "Operational Guidelines" which will include formulas and procedures that are contained in documents different from market practices manual, and (3) "Criteria and Procedures of Operation" which will include specifications, technical notes, and operating criteria required for the implementation of the constituent elements of the market rules in the design of software or daily operations. These protocols are collectively known as "Market Operational Provisions." The Guidelines and Market Operational Provisions together constitute the Market Rules which is equivalent to the tariffs issued by ISO/RTOs in the United States.

Components of WEM to be governed by the market rules include (1) day-ahead market and real-time market for energy and ancillary services, (2) capacity market, (3) market for clean energy certificates, (4) auctions for medium-term energy, (5) auctions for long-term capacity, clean energy, and clean energy certificates, and (6) auctions for financial transmission rights.

The key feature of Mexican WEM is the phased implementation of its various market components. For example, energy and ancillary services market will be implemented in two phases. Phase one will include day-ahead and real-time markets, as well as import/export transactions but exclude demand bidding and virtual bidding. Phase one market is slated to be tested in September 2015 and be fully operational on December 31, 2015. Phase two will include hour-ahead market, demand bidding by controllable resources, and virtual bidding subject to offer validation by market surveillance unit. Both testing and operation of phase two will occur in 2018. Capacity market will have two phases (phase one operational in November 2015 and phase two operational in November 2016).

The general characteristics of Mexican wholesale electricity market are as follows. CENACE will determine the economic dispatch for the entire system after receiving offers/bids from sellers/buyers. Then for each system node, it will calculate the market prices (equivalent to LMPs) which will include system energy price, congestion price, and incremental loss price. Some ancillary services, such as regulation reserves, spinning reserves, and operating reserves, will be competitively supplied through the market while reactive power, and black start will be regulated by CRE.

1.4 ELECTRICITY INDUSTRY RESTRUCTURING IN EUROPE

Much of the European continent is covered by the European Union (EU) which is a unique economic and political partnership among 28 European countries (now 27 after Britain's referendum on June 23, 2016 to leave EU). According to the provisions of European Directive 96/92/EC and then replaced by the Directive 2003/54/EC, the reform of the electricity sector in the EU countries involved the unbundling of electricity generation and trading from the regulated activities of transmission and distribution. Once-monopolistic state-owned electric utilities were broken up to form separate generation and transmission/distribution companies. Competition in the generation sector was introduced while the wires business is still regulated. The primary objectives of industry reform are to establish a feasible and competitive wholesale electricity market which will provide electricity prices to both consumers and generators, guarantee the system security, and ensure an efficient utilization of the resources. The first European countries that embraced the industry reform were England and Wales, and Norway in the 1990s. After that period, almost all European countries have liberalized their power industry and developed their national electricity markets.

EU directives further advocated for gradual opening of national electricity markets to competition, development of common rules for the internal market (generation, transmission, and distribution), non-discriminatory third-party access to transmission networks, and non-discriminatory market-based solutions for cross-border capacity allocation. The directives also encouraged increasing share of electricity to be produced from renewable energy sources and increased trading for greenhouse gas emission allowances.

The electricity market in Europe is a collection of electricity markets in each individual country. Some countries such as Spain and Germany have full-fledged national markets while others are trying to gradually open up similar markets. In a typical national electricity market, the day-ahead market is cleared by matching supply offers and demand bids without considering any network-related constraints. This day-ahead market clearing is typically done by power exchanges. In real-time, the transmission system operators (TSOs) responsible for their respective operating territories schedule the generation according to the day-ahead market outcome. If there is any congestion in real-time by dispatching generation in such a way, the system operators will redispatch generators so as not to overload the system. As such, the system operators also operate the balancing markets (aka real-time markets) in which prices are determined in real-time. These prices form the basis for compensating some generators up for dispatching up and for compensating some generators down for dispatching down. This model of market operation and system operation in European electricity markets is different from the US model in which the market clearing for both day-ahead/real-time markets and system operator functions are jointly done by a single entity. The generic function of the market operator is to facilitate the trading of electricity among sellers, buyers, and traders. This arrangement of electricity trading is generally done by power exchanges in European countries. The goal is to balance demand and supply as well as to discover the market prices for electricity. The reliable operation of the power system is the responsibility of an independent network operator (INO) or TSO.

Power exchanges are auction-based market-clearing mechanisms which generally operate a spot market which includes day-ahead and intraday markets. In the day-ahead auction market, trading of electricity takes place one day ahead for delivery of electricity the next day. Sellers, buyers, and traders of electricity submit their orders electronically, after which supply and demand are compared and the market price is calculated for each hour of the following day. The intraday market can be used to satisfy short-term needs of electricity or to sell short-term overcapacities. Hourly products and flexible block products can be traded. Power exchanges publish the power prices for their own regions and also allow trading of natural gas, coal, CO_2 emission allowances and their derivatives. In some countries, the transmission capacity allocation is managed in the power exchanges in an implicit fashion for congestion management.

The power exchanges in EU electricity markets include EPEX Spot (France, Germany, Austria, and Switzerland), APX (the United Kingdom, Netherlands, and Belgium), Nord Pool Spot (Nordic and Baltic region), OMIE (Iberian Peninsula), Omel (Spain), IPEX (Italy), and PXE (Central Europe). These power exchanges differ to some extent from country to country with respect to market design, regulatory framework, and the background of electricity industry.

EU is intent on pushing for integration of EU national electricity markets. In 2014, day-ahead power markets across 21 European nations were linked through a process known as *market coupling*. Market coupling is a mechanism for integrating markets which allows two or more wholesale electricity market areas (normally corresponding to a national territory) to be merged into a single market area, as long as there are sufficient transmission capacities available among those markets. With market coupling, the daily cross-border transmission capacities among the various areas are not sold separately (*explicitly auctioned*) among the market parties, but is implicitly made available via energy transactions on power exchanges on either side of the border (hence the term *implicit auction*). Buyers and sellers on a power exchange can match their bids and offers submitted via another power exchange as if it was as in one single market area, without the need to separately acquire the corresponding transmission capacities necessary to transport electricity between the two (or more) market areas.

The move was intended to smooth price differences among nations through better control of cross-border power flows. It allows traders to bid for energy on local exchanges, which then automatically allocate cross-border capacity based on price differences with neighbors. It was believed that market coupling can lead to price convergence which can foster more competition. There were also plans to expand that market coupling to intraday markets as well as balancing markets in all EU nations.

Once the market clears and prices are announced, the TSO operates the power system based on the generation schedule determined from the market. For each national market, this step is done by the TSOs in each country. Typically, there is a single TSO for each country except Germany in which there are four TSOs responsible for different regional networks. European TSOs are entities operating independently from the other electricity market players and are responsible for the bulk transmission of electric power on the main high-voltage electric networks. TSOs provide grid access to the electricity market players (i.e., generating companies, traders, suppliers, distributors, and directly connected customers) according to the non-discriminatory and transparent rules. In order to ensure the security of supply, they also guarantee the safe operation and maintenance of the system. In many countries, TSOs are also responsible for the development of the grid infrastructure (system planning). For example, *Swissgrid* is the TSO in Switzerland while *Réseau de Transport d'Électricité* (RTE) is the TSO in France.

For the entire European transmission system, all TSOs form a larger pan-European entity called *European Network of Transmission System Operator for Electricity* (ENTSO-E). ENTSO-E was established and given legal mandates by the EU's third legislative package for the internal energy market in 2009, which aims at further liberalizing the gas and electricity markets in the EU. However, the generation ownership in the EU internal electricity market is still concentrated in Endesa (Spain), Electrabel (Belgium), Vattenfall (Sweden), ENEL (Italy), E.ON and RWE (Germany), and EDF (France). The market design in the EU countries may have to be changed to accommodate the dual goals of energy supply increase from the renewable energy sources and decarbonization of electricity supply.

In the next sections, the development and status of three national markets—the United Kingdom, Nordic, and France—are presented.

1.4.1 The United Kingdom

The United Kingdom is comprised of four countries: England, Wales, Scotland, and Northern Ireland. Up until 1990, the electricity supply industry in England and Wales was under the government ownership. As a vertically integrated monopoly, the Central Electricity Generating Board (CEGB) owns and operates the generation and transmission, and sells electricity to 12 regional area boards (ABs). The regional entities were responsible for distributing and selling electricity to final consumers. In Scotland, there were two vertically integrated Boards: Scottish Power and Scottish and Southern. In Northern Ireland, the Northern Ireland Electricity (NIE) owns and operates an isolated system of transmission and distribution networks.

CEGB was a government-owned utility based on the model of cost-of-service. However, due to the excessive capital costs, high cost of indigenous coal, and low return on assets, the CEGB was restructured and privatized in 1990. It was believed that private ownership and profit motive can provide better incentives than state control approach. The restructuring in England and Wales also represented a model for power sector reform around the world.

Initially, the restructuring involved breaking up CEGB into a separate transmission-only company (National Grid), and three generation companies (National Power, PowerGen, and Nuclear Electric). A power pool was created and generation market was opened for free entry and competition. The market design of the British wholesale electricity market, known as the Power Pool is as follows. The system operator forecasts the demand for each 30-minute period 24 hours ahead. Generator owners submit bid of their choice to the Pool. Then, submitted bids are sorted from the lowest to the highest order and the highest bid needed to just meet the forecasted demand for each 30-minute period sets the pool price, as in uniform-price auction rule. Additionally, successful bidders are paid a capacity charge which can become significant if generation supply is just sufficient to meet the demand. Whole-sale buyers must purchase their electricity requirement from the Pool and pay the Pool price including capacity payment and charges for ancillary services. External to the pool arrangement, bilateral contracts can be struck between any willing buyers and sellers of electricity. In fact, the bilateral contracts accounted for more than 90% of electricity consumption in this pool market design.

While customers with peak loads of more than 1 MW were able to choose their suppliers from 1990, customers with peak load of more than 100 kV were allowed to choose their suppliers from 1994. Since 1999, the remaining part of the electric system (below 100 kV peak load) was opened up for competition. Since that period of privatization, there are more than 30 major power producers operating in Great Britain at the end of 2013.

After privatization, the transmission company (National Grid) was owned by the 12 privatized regional electricity companies, but was floated on the stock exchange in 1995. National Grid has owned and operated the high-voltage transmission system in England and Wales linking generators to distributors and some large customers. The UK transmission system is linked to continental Europe via an interconnector to France under the English Channel. It also has an interconnection with the Netherlands under the North Sea since 2011.

There were some serious concerns about the Power Pool arrangement including high concentration in the generation market, low confidence in the Pool by wholesale buyers for their power purchases, and issue of price signals for the contract markets. For these and other reasons, the British government decided in 1997 to abandon the wholesale market design including the Power Pool that was implemented in 1990. In March 2001, the New Electricity Trading Arrangements (NETA) was introduced in England and Wales to replace the previous Power Pool with new ways to trade electricity. These arrangements were based on the bilateral trading between generators, suppliers, traders, and customers by matching among bids by buyers and sellers via open-access power exchanges (PXs). They were designed to be more efficient and provide greater choices for market participants, while maintaining the operation of a secure and reliable electricity system. The system included forwards and futures markets, a balancing mechanism to enable the National Grid, the transmission system operator, to balance the system, and a settlement process. In the balancing market, the system operator solicits more generator bids if the system-forecast demand is higher than demand bids by the buyers in PX or reduces generation output if the forecast demand is lower than the demand bids.

Up until March 2005, the electricity industry in Scotland, Northern Ireland, and England and Wales was operated independently although all three grid systems are interconnected by transmission ties. Since April 2005, under the British Electricity Trading and Transmission Arrangements (BETTA) introduced under the Energy Act of 2004, the electricity systems of England and Wales and Scotland were integrated. Under this arrangement, National Grid operates a single Great Britain transmission network including the Scottish transmission system.

The electricity supply industry in Northern Ireland has been in private ownership since 1993 with Northern Ireland Electricity (NIE) responsible for power procurement, transmission, distribution, and supply in the Province. Generation is provided by three private sector companies who own the four major power stations. In December 2001, the link between Northern Ireland's grid and that of Scotland was inaugurated. A link between the Northern Ireland grid and that of the Irish Republic was re-established in 1996 to facilitate the transfer of electricity between the two countries. However, on November 1, 2007, the two grids were fully integrated and a joint body "Single Electricity Market Operator (SEMO)" was set up by System Operator for Northern Ireland (SONI) and Eirgrid from the Republic of Ireland to oversee the new single market. In July 2012, an interconnector between the Irish Republic and Wales began operations.

In 1989, a new regulatory office known as the Office of Electricity Regulation (Offer) was formed to regulate the electricity businesses in the United Kingdom. In 2000, it was merged with the Office of Gas Regulation (Ofgas) to form the new Office of Gas and Electricity Markets (Ofgem). Ofgem is an independent national regulatory authority and regulates both gas and electricity industries. The primary role of this office is to protect the interests of existing and future consumers of electricity and gas in the United Kingdom while promoting competition. This regulatory body has a similar role as FERC in the United States.

1.4.2 Nordic Countries

Norway was the first of the Nordic countries to deregulate its power markets. The Energy Act of 1990 formed the basis for deregulation in the other Nordic countries. In 1991, Norwegian Parliament's decision to deregulate the market for trading of electrical energy went into effect. Two years later, *Statnett Marked AS*, a fully owned subsidiary of *Statnett*, was established as an independent company with the goal of providing neutrality and impartiality in operating the electricity market. Total volume in the first operating year was 18.4 TWh, at a value of 1.55 billion Norwegian Krone (NOK). The new framework for an integrated Nordic power market contracts was made to the Norwegian Parliament in 1995. Together with Nord Pool's license for cross-border trading, given by the Norwegian Water Resources and Energy Administration, this framework laid the foundation for spot trading at Nord Pool. A year later, a joint Norwegian–Swedish power exchange was established. The exchange was renamed Nord Pool ASA.

Finland joined Nord Pool ASA in 1998. A year later, *Elbas* was launched as a separate market for balance adjustment in Finland and Sweden. *Elspot* area trade began July 1, the same year. The Nordic market became fully integrated as Denmark joined the exchange in 2000. In 2002, Nord Pool's spot market activities were organized in a separate company, Nord Pool Spot AS. Eastern Denmark joined the *Elbas* market in 2004.

In 2005, Nord Pool Spot opened the Kontek bidding area in Germany, which geographically gives access to the *Vattenfall* Europe Transmission control area. The following year, Nord Pool Spot launched *Elbas* in Germany. The Western Denmark joined the *Elbas* market in 2007. The new *Elspot* trading system, SESAM, was set into production. 2008 saw the highest turnover and market share recorded in the company's history until then. In 2009, Norway joined the *Elbas* intraday market. The European Market Coupling Company relaunched the Danish–German market coupling on November the same year. Nord Pool Spot implemented a negative price floor in *Elspot*.

In 2010, Nord Pool Spot and NASDAQ OMX Commodities launched the UK market *N2EX*. Nord Pool Spot opened a bidding area in Estonia and delivered the technical solution for a new Lithuanian market place. Bidding area was opened in Lithuania by Nord Pool Spot in 2012 while *Elspot* bidding area was opened in Latvia in 2013.

In 2011, *Elbas* was licensed to APX and Belpex as the intraday market in the Netherlands and Belgium, respectively. Intraday market, *Elbas*, was introduced in both Latvia and Lithuania in 2013. Nord Pool Spot took sole ownership of the UK market in 2014. North-Western European power markets were coupled. In 2015, Nord Pool Spot was appointed Nominated Electricity Market Operator (NEMO) across 10 European power markets: Austria, Denmark, Estonia, Finland, France, GB, Latvia, Lithuania, the Netherlands, and Sweden. Nord Pool Spot was rebranded to Nord Pool in 2016. Nord Pool was appointed NEMO in Bulgaria and Germany. Nord Pool has worked with Independent Bulgarian Energy Exchange (IBEX) in opening the Bulgarian day-ahead power market, on January 19, 2016, that will be extended with an intraday market at a later stage. Nord Pool has also worked with Croatian Power Exchange (CROPEX) in launching the day-ahead power market in Croatia on

February 10, 2016. The new CROPEX day-ahead market was operational as a part of the EU-wide multiregional coupling (MRC).

1.4.3 France

The electricity market in France is used to illustrate another example of a national electricity market. In France, the transmission system is owned and operated by French transmission system operator, *Réseau de Transport d'Électricité* (RTE). It also operates a balancing mechanism and cross-border capacity allocation mechanism. In the balancing mechanism, RTE ensures supply–demand balance in real-time, guarantees sufficient operating reserves, and resolves network congestions. France has transmission ties with England (the United Kingdom), Spain, Belgium, Germany, Switzerland, and Italy. Therefore, France is one of the keystone countries for European electricity network and market. Its generation mix is dominated by nuclear technology followed by hydro and coal. Also, France is, primarily, an exporter of electricity to its neighboring countries because of its cheaper supply of electricity.

French electricity market was gradually opened since 1999 with portions of load becoming eligible for participation in the market. In July 2007, all of the loads participated in the entire market. Generation supply is also dominated by *Électricité de France* (EDF) which is mostly owned by the French government. EDF generation made up of about 75–80% of total generation supply in France. The rest of electricity supply came from other competing suppliers. Therefore, the generation market is highly concentrated. In the wholesale market, trading of power is done using bilateral contracts and via power exchanges. The energy industry including both electricity and gas industries, is regulated by an independent regulatory body known as *Commission de Régulation de l'Enérgie* (CRE) which was created in 2000.

1.5 ELECTRICITY INDUSTRY RESTRUCTURING IN ASIA

The most notable electricity market in Asia is the well-developed electricity market in South Korea. Singapore also has a well-developed market while Japan and China are planning to establish such a market.

1.5.1 South Korea

In 1999, the South Korean government set up the "Basic Plan" for the electricity industry restructuring. The basic plan includes implementations of three phases: (1) "Cost-Based Pool" (CBP) (2000–2002), (2) "Two-Way Bidding Pool" (TWBP) (2003–2008), and (3) retail competition (since 2009). In addition to designing the basic plan for the implementation of the CBP market, the operating system for market is designed, and resource scheduling and commitment (RSC) are introduced. The flagship organization, known as Korea Power Exchange (KPX) was established in 2000. Operation of the simulated CBP market was started. At the end of 2000, the Korean government enacted the revised "Electricity Business Act (EBA)" to implement the electricity industry restructuring.

In April 2001, Korean Electricity Regulatory Commission (KOREC) was established under the ministry of commerce, industry and energy (MOCIE), now ministry of trade, industry and energy (MOTIE). The major responsibilities of KOREC include creating a fair and competitive environment for electricity companies by creating standards and rules for the electricity industry, supervising electric companies so that they abide by the rules, resolving disputes among market participants (generator companies and consumers), monitoring and investigating anti-competitive behavior of market participants, enforcing corrective measures for market rule violations, and introducing competition via the restructuring of power industry and review issues related to the rights of electricity consumers.

In 2001, electricity market rules and detailed guidelines were approved by MOCIE. KPX acquired establishment permission and the wholesale electricity market was officially opened on April 2, 2001. The market has the following characteristics that are designed to minimize the restructuring risks that appeared in the initial stages of the market as well as to stimulate more competition.

First, all electricity traders are obligated to participate in the electricity market (pool) in accordance with the EBA's article 31. However, EBA's article 8 allows some exception for these generating companies which have power purchase agreements (PPA) with the Korea Electric Power Corporation (KEPCO). These companies can provide power to KEPCO without trading through the pool. The Act also approves financial contracts for differences (CfD) for market participants who seek to avoid risks.

Second, the Korean electricity market has paid different trading settlements depending on the generator types to stabilize market prices since its opening. The initial market has two sub-markets: "baseload market" and "non-baseload market." The rationale behind this distinction between two such sub-markets was that power generators such as nuclear and coal-fired units can be generally classified into baseload generators which have high fixed costs but low variable costs. On the other hand, generators such as LNG and oil-fired units can be classified into non-baseload generators which have low fixed costs but high variable costs. While baseload generators can maintain stable prices because they are less affected by external factors such as fuel prices and foreign exchange rates, non-baseload generators are more susceptible to such variables. The two-tier pricing system was implemented in 2000 when the baseload plants accounted for 81% of total capacity while the non-baseload generators accounted for the rest. Under such circumstances, pricing based on uniform pricing system (e.g., system marginal price, SMP) could trigger volatility in the wholesale electricity market because prices can swing significantly if non-baseload generators, affected by external factors, set the market price. That can also impact the profitability of generator owners, such as KEPCO and other IPPs.

In 2007, a new form of regulated baseload market price program was introduced in place of abolishing baseload marginal price (BLMP) program. The capacity price of the baseload plants was reduced to the level of non-baseload generators in the same year. In 2008, the regulated baseload market was revised again and thus the two-tier pricing system (baseload and non-baseload) was improved into a single SMP. However, the two-tier pricing structure has been technically maintained by applying the SMP coefficient to the generators which are practically owned by KEPCO, which has over 50% of generator market share. In 2012, the target generators to stabilize market prices started including the coal-fired centrally dispatched generators of private companies. In 2013, the "soft price cap" rule was established to set price cap as the reference price for capacity price and to adjust the settlements by applying the lower price between the market price and the price cap.

Third, generators are mandated to provide ancillary services in accordance with dispatch instructions from KPX. By market rules, scheduled generators must provide ancillary services such as automatic generation control (AGC), governor-free, reasonable reserve margin, reactive power supply, black starts and others which are not compensated during the initial state of market operation. Practical ancillary services settlements were prepared in May 2002, and by readjusting the compensation of governor-free and AGC, actual settlement standards were set up in September 2006. These standards were intended to support the stable operation of power system by properly compensating generators which provide required ancillary services to the system.

Fourth, the operational mode of pumped-storage generators has significant effect on both system and market operations. Thus, the operation of pumped-storage generators has been steadily improved to minimize the operational cost and stabilize the system operation in the market. Consequently, the rules on "price setting scheduling (PSE)" and settlement methods of pumped-storage generators were revised in the electricity market rules in 2011 and 2012. In the new rule, both generation capacity and pumping demand of pumped-storage generators were considered in scheduling PSE. Generators are also encouraged to optimize the operation of their pumped-storage generators. The goal is to minimize the operational cost of the system, strengthen the market price signal and enhance the market efficiency.

Trading payments in the electricity market consists of capacity payments (CP), scheduled-energy-trading payments (SEP), and uplift settlement charges. A SEP is settled in the market for the energy actually generated in accordance with the quota allotted in the PSE. The uplift settlement is the difference between the settlement based on the PSE that does not consider network or generator constraints and the actual settlement made as a result of secure power system operation. It is composed of constrained-on (CON) that is not allocated in the PSE, but generated by power system constraints and constrained-off (COFF) that is allotted in the PSE, but not generated by power system constraints. Capacity payment is made based on the generators' availability declared by power producers until 1 day before the trading day. Capacity price reflects the investment cost and fixed cost of operating the generators. The reference capacity price is annually decided by selecting a standard power plant and calculating the capacity price applicable to that standard power plant.

The Korean electricity market is operated by the following procedures: assessment of power generation costs, system demand forecast, bidding, setting up PSE, determination of SMP, set up of generation schedule, real-time generation including CON and COFF, metering, and settlements. The key feature of Korean electricity market is that it is a CBP where generators are only allowed to bid quantity but not price. Therefore, variable cost curves have to be calculated for setting the market price. As the variable costs cannot be accurately identified in real-time, the actual variable costs are determined a month earlier by assessing the variable cost factors of each power plant.

One day before a trading day, KPX forecasts demand for the next day on hourly basis. Bidding is open until the 10:00 a.m. of the previous day of the trading day. Unlike other electricity markets that allow for bidding both supply quantity and price simultaneously, the bidders in the CBP can bid only the hourly availability of their generators. The PSE is determined based on demand forecast, cost per generator, and availability information of generators to meet system demand at the least-cost. At this stage, only the technical characteristics of power generators are considered while ignoring other constraints such as transmission constraints, heating supply constraints, fuel constraints, and others. Then, the SMP is set by hourly marginal costs based on the PSE. The final SMP is announced by 3:00 p.m. 1 day before the trading day. The real-time operation schedule for the actual power system on the trading day is determined after considering several system constraints such as transmission and fuel constraints while meeting the system demand at the least-cost. On the trading day, power is generated and delivered based on real-time system conditions. Then, the amount of generation by each generator is measured in real-time and settled by the hourly market price.

1.6 RELIABILITY AS A PARAMOUNT GOAL

The single most important purpose of the power system operation is to supply and deliver electricity to final consumers in a reliable manner. At the same time, this must be done in the least-cost manner. This is easier said than done because power system has many components which have complex relationships and intricate dependencies among each other. Sometimes, the degree of interdependencies among these components is such that a failure of a single system element can lead to a failure of the entire system. For example, a failure of a single element can trigger the failures of multiple elements causing a phenomenon known as *cascading failure*. The loss of power in certain portions of the system is called *brownout*. The loss of power triggered by the failure or collapse of the entire power system is known as *blackout*. The blackout is the ultimate system condition that must be avoided at all possible, but not necessarily at all cost.

In the case of a failure of a single system element, this failed element must be removed or isolated from the system as soon as possible before that particular failure triggers the chain reaction to the rest of the system. The goal is to maintain the integrity of the rest of the system. The time needed to remove this failed element can be a few seconds. Automated removal mechanism using sophisticated relays can do this job. After the first isolation, relays will try to reconnect it with the rest of the system after a few seconds. If the failed element cannot be reconnected properly after some attempts, it will be disconnected until the root cause of the problem is identified and resolved by other means.

Unfortunately, any element in the system can fail at any time for any reason. Those elements include a branch of a transmission line, a synchronized generator, a transformer, or other system equipment. System reliability is comprised of two aspects: *security* and *adequacy*. The security of the power system is the ability of the system to withstand a disturbance, large and small, due to failures of some system components. After the disturbance, the system must be restored back to a steady-state condition. Only then, we can say that the system is reliably secure.

Reliable operation of a power system also requires another condition in which the generation resources are sufficiently available. This condition is known as *resource adequacy* or simply *adequacy*. It is a prerequisite condition for the reliable operation of power system. If the available generation resources are not adequate, it would be extremely challenging, if not impossible, to reliably operate the system. Rolling brownouts or blackout can be inevitable for a system with inadequate resources. Such resources can also be a reliable import power from the neighboring systems. In general, import power are less reliable than the generation resources owned and dispatchable by the system. On the other hand, for a system with more than adequate resources, it would be much easier to manage the system and reliable system operation is certainly achievable.

Often times, the status of the transmission network is not given full attention. In fact, having more than enough generation resources is a necessary but insufficient condition. The adequate resources must be complemented by a robust network structure. If the network structure is not robust, power cannot be reliably delivered to final consumers even if there are adequate resources. The weak network structure can cause transmission bottlenecks which will prevent the low-cost power to reach to all final consumers. As a consequence, the reliability of the system can be in peril.

The ever-changing nature of the availability of generation resources and to some extent that of the network structure availability would only make the task of power system operation ever more challenging. This would be a continuous test for the power system operator to get prepared and overcome those challenges so as to deliver electricity to consumers in both reliable and least-cost manners.

So, how much should be spent to achieve an absolutely reliable operation of the system and possibly avoid the blackout at all times? There is no easy answer to this question. If there is any answer, it would largely depend on how much that society is willing to spend to achieve this purpose. On the other hand, is it really worth to spend exorbitant amount of money to achieve a 100% system reliability? In the next section, we will discuss how to achieve this goal via a market mechanism.

1.6.1 Reliability via the Electricity Market Mechanism

How can an electricity market help achieve the goal of reliable operation of a power system? The answer to this question depends on how the market is structured in a particular system. Again, the ultimate goal of the power system operation is to supply and deliver the electricity to final consumers in a reliable manner. How about the cost or economics of supplying that power? The better goal should be the reliable operation of power system in the most economic manner. How can we achieve this?

In a traditional vertically integrated structure, this decision is made by each individual utility which owns the entire supply chain. Under an electricity market structure, the generation sector is open for competition. The underlying theory is that increased economic efficiency can be achieved if the generation sector is exposed to competition and thus, the cost of electricity generation can be lowered. However, with additional structure of electricity market, the complexity of the system also increased because there is one more layer of structure called *electricity market operation*. In a simple term, the function of an electricity market is to solicit offer/bid from generator/load and determine which set of generators should serve the load and hence must be scheduled for specific market periods. As part of this step, market prices are also determined which would become the basis for determining generator revenues and load payments. The key here is the *generation schedule*. A traditional utility also determines this generation schedule for its generators in its own operating area. An electricity market also determines similar generation schedule for its own system. So, what is the difference? The key difference lies in how this set of generation schedule is determined. In a utility setting, it is determined by its own decision-making process without much competitive pressures. In a market setting, it is determined by market forces. Some sort of optimization algorithms are used in both settings to determine that generation schedule, called *optimal generation schedule* or *optimal schedule*.

Once the generation schedule is finalized, the next step is for the system operator to operate the system based on this unique set of generation schedule for specific market periods. It is assumed that competitive pressures and market forces would eventually keep the cost of generation closer to the marginal cost of generation. Only then, we can say that the system is economically short-run efficient because the electricity market is competitive. In fact, electricity markets are responsible only for the cost or economics of scheduling generators, while the system operation takes care of system reliability. Some electricity markets, at least those in the United States, consider and model the network constraints when finding the optimal schedules and determining the market prices. Most power exchanges in Europe do not consider the network constraints which is left to the TSOs to handle. By this way, the system reliability is achieved via an electricity market mechanism.

1.7 FURTHER DISCUSSIONS

For many countries, the restructuring of the electric power system and development of electricity markets are still a work-in-progress. Those countries which have made significant progresses are facing new challenges. These progresses are not without setbacks. Sometimes, external factors, such as national economic condition, can cause significant impact on the successful operation of the electricity market. Many have learned from the trials and errors and have moved forward. In retrospect, no two countries have walked the exact path of restructuring. On the other hand, there are other countries which are seriously considering to restructure their power industries and develop electricity markets. For those countries, there are already many lessons that can be learned. These countries can develop better functioning electricity markets by avoiding some obvious mistakes made by their predecessors.

Based on the free market theory, opening up the generation sector to competition can certainly improve the economic efficiency of the system. However, there are significant costs associated with any major activity such as establishing an electricity market. The benefits, realized or perceived, should outweigh the cost to merit such an endeavor.

The other interesting questions would be: can the market sustain itself? In other words, can a market support a condition in which sufficient number of generators are always available to maintain system reliability? If not, what other options do we have?

CHAPTER END PROBLEMS

1.1   The footprint of the operating electricity markets in the United States does not cover the entire country. Some areas of the country do not have operating electricity markets. What are the drivers for these regions not to pursue the path of developing electricity markets? Are the basis of these drivers related to legal, economic, technical or other issues?

1.2   For power utilities owned by national governments, privatization of that national power company is the first step before the generation sector is open for competition which will eventually lead to full-fledged electricity market. Is this condition strictly necessary for successful restructuring of the industry?

1.3   How can we measure whether the activity of restructuring an electric power industry achieved its stated purpose?

1.4   What are the new challenges faced by the more advanced electricity markets such as those in the United States or Europe? State one challenge and describe the potential solution.

FURTHER READING

1. Hunt S. *Making Competition Work in Electricity*. New York: John Wiley & Sons; 2002.
2. *The Electric Industry in Transition*. Public Utilities Reports, Inc. and the New York State Energy Research and Development Authority; 1994.
3. Plummer JL, Troppmann S. *Competition in Electricity: New Markets and New Structures*. Public Utilities Reports and QED Research; 1990.
4. Lambert JD. *Creating Competitive Power Markets: the PJM Model*. PennWell; 2001.
5. United States Federal Energy Regulatory Commission (FERC) Orders.

Chapter 2

Electric Power System

The purpose of this chapter is to introduce the reader the topic of the electric power system in sufficient detail. For this reason, we will cover the fundamentals of electric power system and its components in this chapter. On a high level, the system components include generator, load, transmission lines, transformers, reactive compensation devices, network model, and distribution system. The authors assume that the reader has some prior knowledge of electrical energy and management, fundamental circuit theories, *Kirchoff*'s laws (both current and voltage laws), and three-phase AC circuits. We also assume that the reader has some familiarity with complex number operation and matrix algebra including algebraic operation and matrix partitioning. Expanded version of this chapter is equivalent to one or two semester's course in electric power engineering major at the undergraduate level.

2.1 ELECTRIC POWER SYSTEM COMPONENTS

Generally, a power system consists of generation, transmission, distribution, and consumption of electric energy. Generating stations include rotating machines (synchronous generators) and other kinds of generators. In generating stations, mechanical or other forms of energy are converted into the electric energy. Electric load typically includes induction motors, lights, and heating elements. For the electric load, the electric energy is converted into mechanical or other useful forms of energy. Transmission lines are represented by either direct current (DC) or three-phase alternating current (AC) system.

The function of a power system is to economically generate electric energy with minimum environmental impact and to transfer this energy over long distance transmission lines and distribution networks with the maximum efficiency and reliability.

Electricity Markets: Theories and Applications, First Edition. Jeremy Lin and Fernando H. Magnago.
© 2017 by The Institute of Electrical and Electronics Engineers, Inc. Published 2017 by John Wiley & Sons, Inc.

That energy is delivered to the final consumers at virtually fixed voltage and frequency in the AC system.

Power system engineering is a discipline rooted in electrical engineering. However, as a power system consists of electrical, mechanical, electronic, and control hardwares, the power engineer needs to have a broad technical knowledge if he is to plan, design, and operate the system as a whole. In addition, he should be aware of the impact of his decisions on the society and the environment.

A power system indeed consists of many more components. The major system components include

1. **Synchronous generators**, which convert mechanical or other types of energy into electrical energy;
2. **Power transformers**, which can step-up or step-down AC voltages with same frequency;
3. **Circuit breakers**, which can open or close the connection of one or several components with other system components under normal operating conditions or abnormal (fault) condition (normal current or overcurrent);
4. **Disconnectors**, which can disconnect system elements;
5. **Instrument transformers**, including voltage transformers and current transformers;
6. **Reactors**, which can limit the overcurrent value during faults;
7. **Surge arrestors**, which protect against the lightning or switching overvoltages.

Generally, immediately after the power is produced from the generator, the voltage is increased by the step-up transformers. Then, this high-voltage power is transmitted via high-voltage transmission lines and continued through subtransmission lines. Before the power reaches the loads, the high-voltage is lowered with the step-down transformers and the power flows through the distribution system and is eventually consumed by the final consumers.

2.2 ELECTRICITY GENERATION

The basic function of a generator is to convert mechanical or other forms of energy into electrical power or electrical energy. Generators can be broadly classified into two types: *rotating machines* and *non-rotating machines*. Rotating machines use turbines which drive the shaft connected to the generator. Non-rotating machines convert one form of energy such as solar energy directly into electrical energy without the need of a turbine. Nearly all of the world's power that is supplied to a major grid is produced by rotating machines with turbines.

Naming conventions for rotating generators are given by the mode by which the turbines are driven. The major types of turbines used in producing electricity are steam turbines, gas-turbines, hydro turbines, and wind turbines. In a steam turbine, the turbine is driven by steam which is produced by burning certain types of fuel. Coal, natural gas, oil, and geothermal sources are used as fuel sources to produce electricity via steam turbines. For example, a steam turbine fueled by coal is called *coal-fired steam-turbine generator*. Generally, the efficiency of geothermal-based power plants is less than that of other steam generators fueled by fossil fuels.

Nuclear power stations use nuclear fuel to produce steam which drives the steam turbine to produce electricity. There are various classifications of nuclear reactors based on the fuel material used, arrangement of elements, and moderator materials, etc. Types of nuclear reactors include pressurized water reactor (PWR), boiling water reactor (BWR), pressurized heavy water reactor (PHWR), basic and advanced gas-cooled reactor (MAGNOX and AGR), and fast breeder reactor (FBR).

Gas-turbines use natural gas which is mixed with compressed air and burned in the combustion chamber to produce high-temperature, high-pressure gases. These hot gases pass through the gas-turbine to drive the generator to produce electricity. In the simple cycle gas-turbine, the exhaust gases which come out from the gas-turbine are directly released into the air. In the combined cycle gas-turbine power plant, the exhaust gases pass through the heat recovery steam generator (HRSG) which reproduces the steam to drive the steam turbine. Because of this additional turbine, the combined cycle power plant has higher efficiency than the simple cycle gas-turbine power plant. The schematics of simple cycle gas-turbine generator and combined cycle gas-turbine generator are shown in Figures 2.1 and 2.2, respectively.

Hydro power plants use falling water to turn the hydro turbine. A hydraulic turbine converts the energy of flowing water into mechanical energy. A hydroelectric generator converts this mechanical energy into electricity. Hydro power plants can start in a few minutes and connect to the grid. And hydro plants are typically housed in dams which have other objectives such as flood control, irrigation, public water

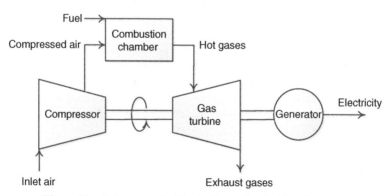

Figure 2.1   Schematic of a Simple Cycle Gas-Turbine Generator.

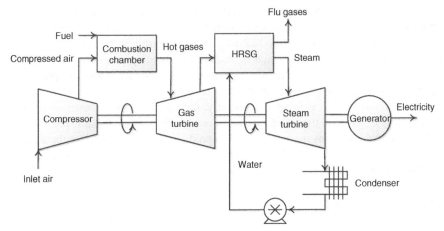

Figure 2.2 Schematic of a Combined Cycle Gas-Turbine Generator.

supply, navigation, and aquaculture. So, generating electricity is not always the primary objective for hydroelectric dams.

Hydroelectric plants are more efficient at providing electricity for peak power demands during short periods than fossil-fuel and nuclear power plants, and one way of doing that is by using *pumped storage*. Pumped storage is a method of keeping water in reserve for peak-period power demands by pumping water that has already flowed through the turbines back up a reservoir above the power plant at a time when customer demand for energy is low, such as during the middle of the night. The water from the reservoir is then released back into the flow through the turbine generators at times when demand is high and a heavy load is placed on the system. Thus, the pumped storage hydro plant acts as both a generator and a load at different times of a day.

Wind turbines use wind energy from the natural environment to produce electricity. The wind turns the blades, which spin a shaft which connects to a generator to generate electricity. Thus, the amount of power output from the wind plants largely depend on the availability of wind in the area. Wind power plants are growing resources in many parts of the world.

Among the non-rotating machines, the solar-powered generator is quite unique because it converts solar energy directly into electrical energy without a rotating machine. Solar-powered generators, such as photovoltaic, are also growing significantly worldwide. Other kinds of generators use tidal energy which spins the turbine to generate electricity and biomass waste which can be burned to generate steam to drive a steam turbine.

2.3 POWER SYSTEM LOAD

The power system load is another major component in the power system. It is for this load that power system planners have to plan for their system with the goal of serving the load reliably. The term *demand* is also used to represent load. System planners are generally concerned about the system's ability to meet the peak demand in the system. Meeting off-peak demand is less of a concern than meeting on-peak demand. Power consumers are generally classified into three broad categories: industrial, commercial, and residential customers. System reliability and power quality are also key requirements in supplying power. Load typically consists of lighting, induction motors, household appliances, air-conditioning, heating, and electronic devices. Different loads require different levels of power quality. For example, the power requirements of typical large computer systems range from 200 to 500 kVA while air-conditioning loads need 25 to 75 kVA. The system operator has to consider real and reactive power supply, voltage and frequency stability, harmonic, and supply continuity. Historically, the size of the generating units and transmission voltage levels increase with the increasing load growth. The growth of load also largely depends on the economic condition of a country or a service territory.

The electrical load in the power system has several unique characteristics. First, electricity must be delivered whenever consumers demand. Electric power, unlike gas and water which can be stored, must be generated whenever it is needed and hence the supplier has little control over the load at any time. The fact of the matter is that in a power system, a sudden demand created by a consumer by switching on a load has to be immediately supplied by an equal infeed of energy at generation points. It follows that the installed generation capacity for a power system should be at least equal to the maximum demand likely to occur on a very hot summer day or on a very cold winter day. For a long period of the year, most of this generation capacity would be idle. This creates a special problem in the electric power system. The control engineers of power system also try to keep the output of electricity from the generators equal to the connected load at the specified voltage and frequency.

Second, the consumers' demand for power varies throughout the day. The consumers' demand on the system is not steady but varies considerably from hour to hour within a day, from day to day within a week, and from season to season. The characteristics of electric energy demand make the electric power supply difficult. It is challenging to meet the daily and yearly load cycles. Power system operators also have to anticipate the unusual events, such as a world cup soccer match or unusually hot summer day, which can cause demand spike.

Third, among the major consumer groups, the power consumption by industrial customers generally accounts for the largest portion of total load in many industrialized countries. The significant electrical device in this load group is the induction motor. Through a combination of examining historical trends and making load forecast for the future, the electric power engineer estimates future generation requirements and recommends facility construction. In addition, the power system planner should also decide the most economical operating state for various loading conditions. This kind of generation planning is typically done in a traditional utility setting. Under a restructured environment with an operating electricity market, the decision to build additional generations is left to the market participants who would decide to build their preferred types of generators based on the market economic signals. The

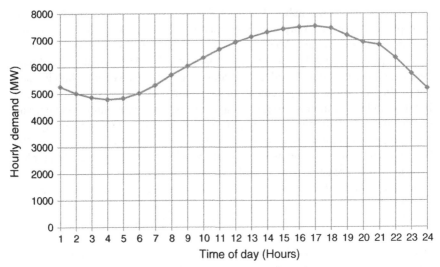

Figure 2.3 Daily Load Curve.

key role of the system planner is to ensure that there is a sufficient level of generation capacity available at any given time in the future.

There are three main factors that greatly influence the magnitude of maximum demand and the time of its occurrence. The most frequent factor is the weather as it affects light intensity during daylight hours and the temperatures throughout the day and the year. The sharpest factor with the shortest duration is special events which can result in a temporary slowdown of activities or sudden jump of demand due to increased usage of lighting, radio, TV, water pumping, cooking, and other load increases. The largest factor is the changes in business conditions accompanied by significant changes in industrial demands and consumptions. The changes in demand in both commercial and residential consumers are much less significant. The nature, magnitude, and time of these fluctuations are generally unpredictable.

The load curve or demand curve is a chart that shows the variation of load with respect to time. The daily load curve is the load that varies each hour in a day while the yearly load curve shows the varying load for each month in a year. A typical daily load curve and a weekly load curve are shown in Figures 2.3 and 2.4, respectively. Hourly loads must also be estimated in order to economically meet each day's demand. The functions of a load curve are (1) to show the regularity of load change, (2) to calculate the electricity consumption of customers, (3) to forecast the daily or yearly load consumption, and (4) to dispatch the generation output according to the daily load curve.

To meet the peak demand, sometimes an additional set of generators must be turned on for only a few hours even if it is less efficient and more expensive to operate. Generating units with quick-start capability are suitable for this purpose.

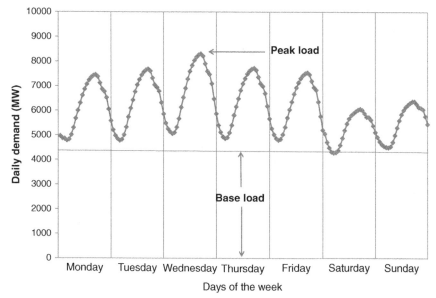

Figure 2.4 Weekly Load Curve.

Therefore, gas-turbine generators fit this category. Just as the load varies each hour in a day, the peak load of each day also varies throughout the year. The yearly load variation is very important in scheduling maintenance of equipment especially large generators. During peak load periods, it is desirable to have all the equipment available online to be able to meet the system demands. The power engineer must consider the necessity of maintenance in determining when to take the equipment out of service.

In the load curve, the portion of load that does not change with time is called *base load*. The rest of the upper portion of the load is called *peak load* or the combination of peak load and *intermediate load*. Nuclear plants and other steam-turbine generators are good candidates for meeting base load because their output efficiency is at the highest when running at full load. On the other hand, gas-turbine generators are suitable for meeting peak load because they have a quick-start capability. The area under the load curve is the total energy demanded of the system for a given duration.

2.4 TRANSMISSION LINES

The network of transmission lines is the major nervous system of a power system. The key function of transmission lines in a network is to connect power generating stations to the load centers through the transformers by enabling the transfer of power. Transmission lines can be classified into three broad categories:

1. Transmission system (typically 345 kV and above)
2. Subtransmission system (between 35 kV and 345 kV)
3. Distribution network (35 kV and lower)

An overhead transmission line mainly consists of conductors, insulators, support structures (towers), and shield wires (earth grounding wires). All transmission lines in a power system exhibit the electrical properties of resistance, inductance, capacitance, and conductance.

The benefits of using high-voltage when transmitting power include reduced energy loss on the lines, reduced conductor cost, and the ability to transfer power over the long distance. However, with high-voltage power transfer, the costs of transmission lines, components, and substations also increase due to higher towers, insulators, and increased insulations for each component. Generally, the categories of high-voltage consist of

1. High-voltage (HV) (1 kV–220 kV)
2. Extra-high voltage (EHV) (330 kV–500 kV)
3. Ultra-high voltage (UHV) (750 kV and above)

The voltage condition of a power system, particularly at various bus locations, is also very important for reliable system operation. A voltage profile of a typical power system is shown in Figure 2.5. Generally, the voltage at the end-use location should not be lower than 95% of the rated voltage of the consumed equipment. The voltage drop of the transmission lines should not be more than 10%. Therefore, the voltage at the supply side should be higher than 5% of the rated voltage of the consumed equipment.

2.4.1 Structures and Types of Transmission Lines

There are many types of transmission lines that have been used today. Many factors affect the selection of transmission structures as well. It is impossible to say one

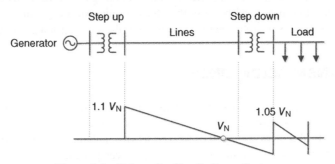

Figure 2.5 Voltage Profile of a Power System.

type is better than the other. The choice of the transmission line design depends on the optimization of electrical, mechanical, environmental, and economic factors. An overhead transmission line consists of

1. **Conductors**: Aluminum has become the most common conductor metal for an overhead transmission line due to the lower cost, lighter weight, and abundant resources. The most common conductor type is aluminum conductor, steel-reinforced (ACSR). Stranded conductors are easier to manufacture, handle, and are more flexible. The steel strands have a high strength-to-weight ratio. EHV lines normally have a bundled subconductor design. In this design, each of the three phases consists of two, three, or four separate conductors per phase. Bundle conductors have a lower electric strength at the conductor surfaces to control the corona and also have a smaller series reactance.

2. **Insulators**: The function of an insulator is to isolate the energized conductors from ground (tower). Insulators for transmission lines above 35– 750 kV are suspension-type insulators which consist of a string of discs typically made of porcelain. Glass and composite polymeric materials are also used. The number of insulator discs in a string increases with increasing line voltage.

3. **Tower Structures**: Categories of tower materials and types of tower include lattice steel towers, tubular steel towers, wood poles, and concrete poles. Configuration of tower can be vertical, triangular, and horizontal.

4. **Shield Wires (Earth Grounding Wires)**: The function of shield wires is to prevent the transmission lines from direct lightning strike. Shield wires located above the phase conductors protect the phase conductors against the lightning. Numbers and locations of the shield wires are selected in such a way that almost all lightning strikes terminate on the shield wires rather than on the phase conductors. The angle between shield wire and line is about 20–30 degrees. Shield wires act as ground to the tower to provide the tower impedance and tower footing resistances are small.

2.4.2 Electrical Parameters of Transmission Lines

All transmission lines in a power system exhibit unique electrical properties with four basic parameters: (1) series resistance which accounts for ohmic (I^2R) line losses, (2) series impedance including resistance and inductance which gives rise to the series-voltage drops along the line, (3) shunt capacitance which gives rise to line-charging currents, and (4) shunt conductance which accounts for (V^2G) line losses due to leakage currents between conductors or between conductors and ground. Shunt conductance of overhead lines is usually neglected.

The fundamental parameters of transmission lines: resistance (R), conductance (G), inductance (L), and capacitance (C) are further explained.

1. **Resistance of Transmission Lines**: The series resistance of a transmission line is affected by the resistivity (ρ) of its conductors, spiraling of the strands within the conductors, temperature, and skin effect. For stranded conductors, alternate layers of strands are spiraled in opposite direction to hold the strands together because the strands must be longer than the actual length of the line in order to spiral them. Therefore, the resistance of the transmission line is greater than the resistance due to the actual length of the line. The extra length increases resistance by 1–2%. Also, the effective cross-sectional area of the conductor is reduced due to the skin effect and its effective resistance at 60 (50) Hz is a few percent higher than its DC value. For DC, the current distribution is uniform throughout the conductor cross section and ($R = (\rho \times l)/A$) (where l = conductor length, A = cross-sectional area of the conductor) is valid. However, for AC, the current distribution is nonuniform. As frequency increases, the current in a solid cylindrical conductor tends to crowd toward the conductor surface with smaller current density at the conductor center. This phenomenon is called *skin effect*. Skin effect is a function of conductor size, frequency, and the relative resistance of the conductor material.

2. **Conductance of Transmission Lines**: The conductance accounts for real power loss between conductors or between conductor and ground. For transmission lines, this power loss is due to leakage currents at insulators and due to corona. Insulator leakage current depends on the amount of dirt, salt, and other contaminants that have accumulated on insulators as well as on meteorological factors, particularly the presence of moisture. Corona occurs when a high value of electric field strength at a conductor surface causes the air to become electrically ionized and to conduct. The real power loss due to corona, called *corona loss*, depends on the meteorological conditions, particularly fog and on conductor surface irregularities. Losses due to insulator leakage and corona are usually small compared to conductor (I^2R) loss. Conductance is usually neglected in power system studies because it is a very small component of the shunt admittance.

3. **Inductance of Transmission Lines**: The inductance of a transmission line consists of two components—self-inductance and mutual inductance. The self-inductance also has two components—internal and external inductance. The inductance of a magnetic circuit that has a constant permeability (μ) can be obtained by determining magnetic field intensity, magnetic flux density, flux linkages, and inductance from flux linkages per ampere. Series inductance of a transmission line can be defined in terms of the flux linkage of its conductors. The internal conductor inductance is due only to the current flow in the conductor per unit length and is independent of conductor radius. Each

conductor of a transmission line has inductance due to external magnetic flux linkage.

4. **Capacitance of Transmission Lines**: Transmission line conductors exhibit capacitance with respect to each other owing to the potential difference between them. The amount of capacitance between conductors is a function of conductor size, spacing, height above ground, and voltage. Capacitance of a conductor with respect to any other point is defined as $(C = q/v)$ (where q = instantaneous charge on the conductor and v = voltage drop from the conductor to a point of interest). For single-phase transmission lines, some of the nonuniform charge distribution is caused by the close proximity of conductors a and b. However, this distortion has a negligible effect on capacitance calculations and can be ignored. The most significant distortion is caused by the effect of the ground on the conductors. If the height of conductors is comparatively higher than the conductor spacing, which is a standard for most transmission lines, the capacitive effect of the ground on the conductors can also be neglected. Capacitance calculations for three-phase transmission lines are usually performed by calculating the line-to-neutral capacitance values for each conductor. This requires solving for the line-to-neutral voltage of each conductor as a function of the conductor's charge.

2.4.3 Electric Network Models of Transmission Lines

Analysis of a power system requires a mathematical model of transmission lines in order to calculate voltage, current, and power flows. This model must account for the series resistance, series inductance, and shunt capacitance of each phase. These quantities are distributed along the entire length of the line. Calculations of voltages and currents at any point on the line can be done as follows with reference to Figure 2.6.

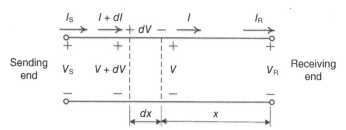

Figure 2.6   One Phase of a Transmission Line.

At a distance x meters from the receiving end, a voltage drop dV is caused by a current flow I through the series impedance:

$$dV = I\,zdx \quad \text{or} \quad \frac{dV}{dx} = Iz \tag{2.1}$$

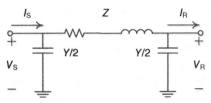

Figure 2.7   Equivalent π Circuit of One Phase of a Transmission Line.

and

$$dI = Vydx \quad \text{or} \quad \frac{dI}{dx} = Vy, \tag{2.2}$$

where the series impedance $z = r + j\omega L$ (Ω/m) and the shunt admittance $y = j\omega C$ (mho/m), r = series resistance (Ω/m), ω = angular frequency of the voltages and currents (rad/sec), L = series inductance (H/m), and C = line-to-neutral or shunt capacitance (F/m). And,

$$\frac{d^2V}{dx^2} = yzV \quad \text{and} \quad \frac{d^2I}{dx^2} = yzI \tag{2.3}$$

Equivalent π circuit of one phase of a transmission line is shown in Figure 2.7. The boundary conditions are $V_{x=0} = V_R$ and $I_{x=0} = I_R$. The results are

$$V(x) = \frac{1}{2}(V_R + I_R R_c)e^{\gamma x} + \frac{1}{2}(V_R - I_R Z_c)e^{-\gamma x} \tag{2.4}$$

$$I(x) = \frac{1}{2}\left(\frac{V_R}{Z_c} + I_R\right)e^{\gamma x} - \frac{1}{2}\left(\frac{V_R}{Z_c} - I_R\right)e^{-\gamma x}, \tag{2.5}$$

where $Z_c = \sqrt{\frac{z}{y}}$ is the characteristic impedance of the line and $\gamma = \sqrt{zy}$ is the propagation constant. For a transmission line that is l meters long, Z and Y are calculated as

$$Z = Z_c \ \sinh(\gamma l) \quad \text{and} \quad Y = \frac{2}{Z_c} \ \tanh\left(\frac{\gamma l}{2}\right) \tag{2.6}$$

For transmission lines longer than 250 km, Eq. (2.6) should be used. For values of l in the range of 80–250 km, the Z and Y values can be approximated as

$$Z = zl \quad \text{and} \quad Y = yl \tag{2.7}$$

For transmission lines shorter than 80 km, the value of Z can still be calculated using Eq. (2.7). However, the capacitance and leakage resistance to the earth are

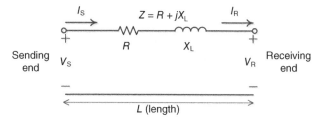

Figure 2.8   Circuit Representation for a Short Transmission Line.

usually neglected because Y is small enough when compared with Z. Therefore, the transmission line can be treated as a simple, lumped, and constant impedance, that is,

$$Z = R + jX_L = zl = rl + jxl, \tag{2.8}$$

where Z = total series impedance per phase (Ω), z = series impedance of one conductor (Ω) per unit length, X_L = total inductive reactance of one conductor (Ω), r = series resistance of one conductor (Ω) per unit length, x = inductive reactance of one conductor (Ω) per unit length, and l = length of the line.

The current entering the line at the sending end of the line I_S is equal to the current leaving at the receiving end I_R. Figure 2.8 shows the vector (or phasor) diagram for a short transmission line connected to an inductive load. It can be observed from the figure that

$$V_S = V_R + I_R Z \quad \text{and} \quad I_S = I_R = 0V_R + I_R \tag{2.9}$$

or

$$V_R = V_S - ZI_S \quad \text{and} \quad I_R = I_S = 0V_S + I_S \tag{2.10}$$

The generalized circuit constants or *ABCD* parameters, can be determined by the inspection of the figure. Since

$$\begin{bmatrix} V_S \\ I_S \end{bmatrix} = \begin{bmatrix} A & B \\ C & D \end{bmatrix} \begin{bmatrix} V_R \\ I_R \end{bmatrix} \tag{2.11}$$

and $AD - BC = 1$, where $A = 1$, $B = Z$, $C = 0$, $D = 1$, then,

$$\begin{bmatrix} V_S \\ I_S \end{bmatrix} = \begin{bmatrix} 1 & Z \\ 0 & 1 \end{bmatrix} \begin{bmatrix} V_R \\ I_R \end{bmatrix} \tag{2.12}$$

and

$$
\begin{bmatrix} V_R \\ I_R \end{bmatrix} = \begin{bmatrix} 1 & Z \\ 0 & 1 \end{bmatrix}^{-1} \begin{bmatrix} V_S \\ I_S \end{bmatrix} = \begin{bmatrix} 1 & -Z \\ 0 & 1 \end{bmatrix} \begin{bmatrix} V_S \\ I_S \end{bmatrix}
\tag{2.13}
$$

2.5 POWER TRANSFORMERS

The power transformer is a major power system component that permits economical power transmission with high efficiency and low series-voltage drops. Power transformers transform AC voltage and current to optimal levels for generation, transmission, distribution, and utilization of electric power.

The function of a transformer in a power system is to step the terminal voltage of the generator up to the voltage of the transmission system and at the terminal of the transmission line to step the voltage down to the customers or substations. The reason for stepping up the voltage is to reduce I^2R losses within the transmission system. Today's modern power transformers have nearly 100% efficiency with ratings up to and beyond 1300 MVA. The three-phase power that is transmitted on a transmission line is

$$
P = \sqrt{3} V_{ll} I_1 \cos \theta
\tag{2.14}
$$

where V_{ll} is line-to-line voltage, I_1 is line current, and θ is the voltage angle difference between sending end and receiving end. The I_1 will have an inverse relation with V_{ll} if the power factor and the power being transmitted remain constant and raising V_{ll} will reduce I_1. Therefore, for line loss consideration and reduction of voltage drop along the line, higher voltages are desirable for transmission of power over long distances.

The configuration of power transformers includes winding (primary and secondary or low-voltage and high-voltage coil), core, bushing (high-voltage and low-voltage bushing), and cooling systems (self-cooled, forced air-cooled, and forced oil-cooled). Nameplate of a transformer is determined by rated voltage (primary and secondary) (kV), rated current (A), rated voltampere (MVA), and impedance (%). Basic parameters of a transformer include magnetic flux, magnetic flux density, magnetic field intensity, flux linkage, and permeability (μ). The power losses in a power transformer has two components: core loss and copper loss. The core loss also consists of eddy current losses and the hysteresis losses, both of which are real-power loss.

In a three-phase transformer, all six windings are on a single core in one tank. A three-phase transformer is usually either a core or a shell type. Three-phase transformer ratings are always in line-to-line voltage and three-phase MVA. The advantages of a three-phase transformer include lower cost, less space requirements, and lower weight. A three-phase transformer can be connected in any of the connections: Y–Y, delta–delta, and Y–delta or delta–Y connections.

An autotransformer is one in which one winding serves both primary and secondary functions. Autotransformers are increasingly used to interconnect two high-voltage transmission lines operating at different voltages. For the autotransformer, the windings are both electrically and magnetically coupled. Advantages of an autotransformer include (1) less copper for the same output, (2) the greater efficiency for the same output, (3) the superior voltage regulation, and (4) smaller in size for the same output.

2.6 SYNCHRONOUS GENERATORS

Synchronous generators are universally used in the electric power industry for supplying three-phase as well as single-phase power to the electricity consumers. Synchronous generators are classified into two types. The first is the low-speed (water-driven) type which is characterized physically by having salient poles, large diameter, and small axial length. The second is the turbo-generator (cylindrical rotor) which uses the steam turbine as the prime mover which has non-salient poles, relatively smaller diameter, and longer axial length with high speed during the operation.

There are two principal parts in a synchronous generator: the *stator* (stationary part) and the *rotor* (rotating part). The stationary part which is essentially a hollow cylinder is called the *armature* and has longitudinal slots in which coils of the armature winding are wound. This armature winding carries the current supplied to an electrical load or system. The rotor is the other part of the generator which is mounted on the shaft and rotates inside the hollow stator. The winding on the rotor is called the *field winding* and is supplied with DC current. The very high magnetomotive force (MMF) produced by this DC current in the field winding combines with the MMF produced by current in the armature winding. The resultant flux generates voltage in the coils of the armature winding and provides the electromagnetic torque between the stator and the rotor.

The excitation current (field current) is produced by an adjustable DC source, such as a self-excited DC generator, or else through electronic rectification. In the rotor, the field winding is distributed in the manner in order to produce a nearly sinusoidal distribution of flux density in the air gap. The slotted stator core is composed of laminations of high-grade sheet steel to provide a low-reluctance magnetic path for the operating flux produced by the rotor field winding. The stator has three sets of windings or phases labeled *a*, *b*, and *c* (for a three-phase generator). The windings are centered 120 degrees apart around the inner surface of the stator. Each winding provides one phase of the voltage required for the three-phase loads.

Equivalent electric circuit of a generator is necessary for the analysis of power system. The circuit is constructed by developing relationships between the terminal voltages of the stator windings and the currents of the stator and the rotor. Applying the *Faraday's law* to the coil *a* shows that the value for V_a is

$$V_a = V_{ar} + V_f, \tag{2.15}$$

where $V_{ar} = 3\omega_s L_s I_{max} \sin(\omega_s t)$ is the armature reaction voltage, $V_f = \omega_s L_{sr} I_r \sin(\omega_s t + \gamma)$ is the field excitation voltage (the voltage induced on the stator windings under no-load conditions), and V_{ar} represents the effect of MMF created by current flowing in the stator on the resultant MMF. Then, the equivalent circuit of a synchronous generator can be expressed as

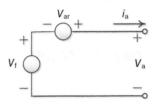

Figure 2.9   Circuit Model of Phase a using V_f and V_{ar}.

2.7 NETWORK ANALYSIS

Successful operation of the power system under a normal, balanced three-phase steady-state condition requires the following conditions:

1. Generation output equals the demand (load) plus losses
2. Bus voltage magnitudes remain close to rated values
3. Generators operate within specified real and reactive power limits
4. Transmission lines and transformers are not overloaded

The primary objective of investigating the power flow (aka load flow) for a power system is to estimate the key parameters of voltage, real and reactive power flow, and power losses. Power flow studies are performed to investigate the following:

1. Flow of real and reactive power in the branches of the network
2. Busbar voltages
3. Effect of rearranging circuits and incorporating new circuits on the system loading (expansion)
4. Effect of temporary loss of generation and transmission circuits on system loading
5. Optimal system running conditions and load disturbance
6. Optimal system losses
7. Optimal rating and tap range of transformers
8. Improvement from change of conductor size and system voltage

Conventional nodal or loop analysis is not suitable for power flow studies because the input data for loads are normally given in terms of power, not impedance. Also, generators are considered as power sources, not voltage or current sources. Power flow problem is formulated as a set of nonlinear algebraic equations and the computer software program is typically used to solve that problem.

2.7.1 Impedance Diagram

A balanced three-phase system is almost always analyzed as a single-phase circuit for simplicity. This kind of analysis means that the system is represented by a diagram of one phase with an assumed neutral return. A single-phase representation for balanced three-phase operation may be used because the electrical parameters calculated for one phase will be exactly the same in the other two phases (with a ± 120 degree phase shift). A diagram of a network is simplified by representing its elements by symbols rather than by their equivalent circuits. Such a diagram is called *one-line diagram* or *impedance diagram*. For calculations, the symbols of the one-line diagram must be replaced by single-phase equivalent circuits of the elements. The per-phase impedance diagram may present a system given in *ohms* or in *per units*. For such an equivalent impedance diagram, several rules are normally specified:

1. The buses (i.e., the nodal points of the transmission network) have been identified by their bus numbers and their resistances are neglected.
2. Assuming that the transformer magnetization currents are negligible and then can be represented as series impedances.
3. Transmission lines are represented as straight lines and short. They may be represented as a resistance and inductive reactance in series.
4. The neutral of generator has either been solidly grounded or grounded through impedance or resistance. They are used to limit the current flow to ground under fault condition. Synchronous generator symbols can be replaced with impedances and voltage sources.

Notice that the impedances of the grounding circuits of the generators are not included in the impedance diagram. Under balanced three-phase conditions, no current would flow in the neutral, so the neutral impedance would have no effect on the system. Figure 2.10 shows an example of a one-line diagram, and the relevant equivalent impedance diagram per phase.

2.7.2 Bus Admittance Matrix

A simple power system is shown as one-line diagram in Figure 2.11a. If the system impedances are known, an equivalent impedance diagram described in the

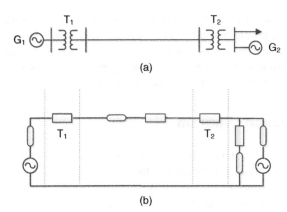

Figure 2.10 (a) One-Line Diagram (b) Equivalent Impedance Diagram per Phase.

previous section can be drawn. However, using admittances (Y) is more convenient for writing nodal equations. Figure 2.11b shows the admittance diagram of the given one-line diagram. These admittances are reciprocals of their corresponding impedances.

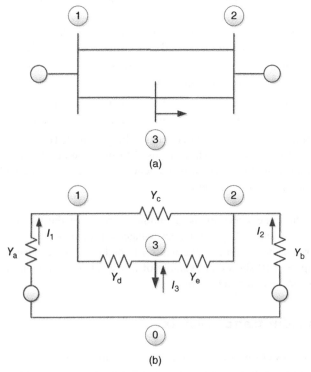

Figure 2.11 (a) One-Line Diagram (b) Associated Admittance Diagram.

The nodal equation for the system is as follows. Since the current entering a node from a generator or load is equal to the current leaving that node through all other system elements, the nodal expression for node 1 is

$$I_1 = (V_1 - V_2)Y_c + (V_1 - V_3)Y_d \qquad (2.16)$$

where I_1 is the current entering node 1 from the generator at node 1. For node 2 and node 3, similar expressions can be written as

$$I_2 = (V_2 - V_1)Y_c + (V_2 - V_3)Y_e \qquad (2.17)$$

$$I_3 = (V_3 - V_1)Y_d + (V_3 - V_2)Y_e \qquad (2.18)$$

These are three independent equations with three unknowns. If I_1, I_2, and I_3 are known or can be calculated, a solution for values of V_1, V_2, and V_3 can be found and using these voltages, all branch currents can be determined. This conclusion implies that the number of required nodal equations needed to solve any circuit is one less than the number of nodes in the circuit. Rearranging the above questions:

$$\begin{aligned}
I_1 &= (Y_c + Y_d)V_1 + (-Y_c)V_2 + (-Y_d)V_3 \\
I_2 &= (-Y_c)V_1 + (Y_c + Y_e)V_2 + (-Y_e)V_3 \\
I_3 &= (-Y_d)V_1 + (-Y_e)V_2 + (Y_d + Y_e)V_3
\end{aligned} \qquad (2.19)$$

These equations can be written in the following form:

$$\begin{aligned}
I_1 &= Y_{11}V_1 + Y_{12}V_2 + Y_{13}V_3 \\
I_2 &= Y_{21}V_1 + Y_{22}V_2 + Y_{23}V_3 \\
I_3 &= Y_{31}V_1 + Y_{32}V_2 + Y_{33}V_3
\end{aligned} \qquad (2.20)$$

The admittances Y_{11}, Y_{22}, and Y_{33} are self-admittances of each node and equal to the sum of all admittances except generators and load connected to that node. The admittances $Y_{12}, Y_{13}, Y_{21}, Y_{23}, Y_{31}$, and Y_{32}, are mutual admittances. The mutual admittances are equal to the negative of the sum of the admittances directly connected to the nodes identified by their subscripts. Written in the matrix form:

$$\begin{bmatrix} I_1 \\ I_2 \\ I_3 \end{bmatrix} = \begin{bmatrix} Y_{11} & Y_{12} & Y_{13} \\ Y_{21} & Y_{22} & Y_{23} \\ Y_{31} & Y_{32} & Y_{33} \end{bmatrix} \begin{bmatrix} V_1 \\ V_2 \\ V_3 \end{bmatrix} \qquad (2.21)$$

The square matrix of self and mutual admittances in Eq. (2.21) is called the *bus admittance matrix.*

2.7.3 Power Flow Analysis

Given the terminal voltage magnitudes, angles, and the impedances of the transmission line model of Figure 2.12, the flow of real and reactive power from one terminal toward the other terminal can be calculated. The phase voltages at each terminal will be

$$V_S = V_1 \underline{/\theta_1} \quad \text{(terminal 1)}$$
$$V_R = V_2 \underline{/\theta_2} \quad \text{(terminal 2)}$$

(2.22)

The series impedance Z may also be represented as an admittance Y:

$$\frac{1}{Z_{12}} = Y_{12} = g - jb \tag{2.23}$$

The complex power flowing from terminal 1 toward terminal 2 is the summation of the flows in the shunt branch at terminal 1 and the series branch connecting terminals 1 and 2. The complex power flow in the shunt branch is

$$S_{S1} = V_1 \underline{/\theta_1} \left[(Y/2)(V_1 \underline{/\theta_1}) \right]^* = V_1^2 (Y/2)^* \tag{2.24}$$

Since conductance is neglected for power transmission lines, S_{S1} is composed of reactive power only. The complex power flow in the series branch is

$$S_{12} = V_1 \underline{/\theta_1} \left[(V_1 \underline{/\theta_1} - V_2 \underline{/\theta_2}) Y_{12} \right]^* \tag{2.25}$$

or

$$S_{12}^* = V_1 \underline{/-\theta_1} [(V_1 \underline{/\theta_1} - V_2 \underline{/\theta_2})(g - jb)] \tag{2.26}$$

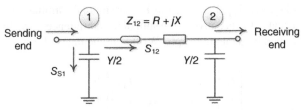

Figure 2.12 Equivalent Circuit of One-Line Diagram of Transmission Line.

$$S^*_{12} = \left[V_1^2 g - V_1 V_2 g \cos (\theta_2 - \theta_1) - V_1 V_2 b \sin (\theta_2 - \theta_1) \right]$$
$$+ j \left[- V_1^2 b + V_1 V_2 b \cos (\theta_2 - \theta_1) - V_1 V_2 g \sin (\theta_2 - \theta_1) \right] \quad (2.27)$$

then,

$$P_{12} = V_1^2 g - V_1 V_2 g \cos (\theta_2 - \theta_1) - V_1 V_2 b \sin (\theta_2 - \theta_1)$$
$$Q_{12} = V_1^2 b - V_1 V_2 b \cos (\theta_2 - \theta_1) + V_1 V_2 g \sin (\theta_2 - \theta_1) \quad (2.28)$$

When the angle difference between θ_1 and θ_2 is small, the following approximations can be made:

$$\cos (\theta_2 - \theta_1) \approx 1.0$$
$$\sin (\theta_2 - \theta_1) \approx (\theta_2 - \theta_1) \quad (2.29)$$

Also since the value of series inductive reactance for a transmission line is normally much greater than the series resistance, b will be much greater than g. Then,

$$P_{12} = V_1 V_2 b (\theta_1 - \theta_2)$$
$$Q_{12} = V_1 b (V_1 - V_2) \quad (2.30)$$

Two important conclusions can be drawn here. First, the flow of real power (P) on a transmission line is determined by the angle difference of the terminal voltages. Second, the flow of reactive power (Q) is determined by the magnitude difference of terminal voltages.

In power flow studies, classification of buses is also important. There are mainly three types of buses:

1. **Load Buses**: No generation is present in these buses. The demanded real and reactive powers of the load are known. Load buses are also called *PQ bus*.
2. **Voltage Control Buses**: These are generator buses where voltage magnitude and real power being generated are known. In some cases, generator buses may be treated as load buses if a reactive power generation is specified at that bus instead of a voltage magnitude. Voltage control buses are also called *PV bus*.
3. **Swing Bus**: This bus has a specified voltage magnitude and angle. All other bus voltage angles are referenced to it. The swing bus serves the function of supplying or demanding power to or from the system to keep the total system power generated and consumed including system losses in balance.

Calculation of bus voltages requires values for the injected currents. These injected currents can be calculated if certain other parameters are known. Given the complex power (S) at bus i as

$$S = P_i + jQ_i = V_i I_i^*$$
$$S^* = P_i - jQ_i = I_i V_i^*$$

(2.31)

then,

$$I_i = \frac{P_i - jQ_i}{V_i^*}$$

(2.32)

The objective of running a power flow analysis is to determine values for any unknown parameters at each of these three types of buses. For simplicity, Table 2.1 shows which parameters are known and which are calculated from the power flow analysis. Those parameters to be calculated from the power flow analysis are marked with x. Once the power flow analysis is done (assuming convergence is reached within a reasonable tolerance), we would know each of the four parameters at every node (bus) of the entire network system. The available iterative methods to solve load flow include Gauss–Seidel method, Newton, Newton–Raphson (full or fast decoupled), three-phase backward/forward, and DC load flow method in which both V and Q are ignored.

Table 2.1 Parameter Identifications of Three Bus Types

Bus Type	V	δ	Net P	Net Q
Load bus	x	x	Known	Known
Voltage control bus	Known	x	Known	x
Swing bus	Known	Known	x	x

2.7.4 Control of Power Flow

The results of the power flow analysis make it clear that controls of power flow are necessary to ensure the attainment of specified voltage levels as well as specific values of real and reactive power at each bus in the system. The following means are used to control system power flows.

1. **Prime Mover and Excitation Control of Generators:** Assume that the generator to be discussed is part of a large power system. The system is so large and strong that every bus in the system appears to be an infinite bus.

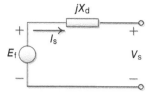

Figure 2.13   Equivalent Circuit Diagram of a Synchronous Generator.

That means that the voltage at the generator bus will not be altered by the changes in the generator's operating conditions.

From the equivalent diagram for that generator shown in Figure 2.13, the following equations can be obtained:

$$P = \frac{V_s E_f}{X_d} \sin\delta$$

$$Q = \frac{V_s}{X_d}(E_f \cos\delta - V_s)$$

(2.33)

From this set of equations, it can be seen that if V_s and E_f are held constant (X_d is also constant), the real power output is directly dependent on the voltage angle δ. In a steady-state operation, the real power output can only be increased by an increase of real mechanical power input. Therefore, δ must also be directly dependent on the mechanical power input.

The equation also shows that if V_s and E_f are held constant, Q would appear to be affected by changes in δ or changes in mechanical power input. However, the normal operating range for δ is less than 15 degrees. For small changes of 15 degrees in this operating range, the *cosine* term is relatively unchanged while the *sine* term will vary significantly. Therefore, the real power P output is directly dependent on the changes in the mechanical power input.

Figure 2.14 shows the phasor diagrams for a generator supplying reactive power to the system where (a) is overexcited, (b) in-phase, and (c) is underexcited. If the real-power output is held constant while the DC excitation on the rotor is varied, an interesting phenomenon occurs. Since the real-power output is constant, the projection of I_s onto V_s must remain constant. Therefore, if the magnitude of E_f with the rotor DC excitation is changed, the magnitude of I_s and the angles θ and δ also change. The reactive power output of a generator appears to be controlled by adjusting the rotor DC excitation. By controlling the amount of real and reactive power generated at each generating station in a system, the general flow of power in the system can be controlled.

2. **Switching of Shunt Capacitor Banks and Shunt Reactors:** From the complex power, only the real power that is delivered to the load actually does real mechanical work. However, most of the loads in a power system are inductive.

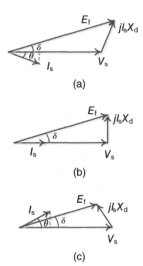

Figure 2.14 Phasor Diagrams of a Generator Supplying Reactive Power (a) Overexcited (b) In-Phase (c) Underexcited.

Therefore, reactive power must also be generated to supply this inductive load. In fact, delivering the same amount of real power to a load with a reactive part requires more current than supplying load with no reactive part. Larger currents mean greater voltage drops along transmission and distribution lines and thus more I^2R line losses.

One way to overcome this problem is to generate reactive power at the location where inductive load is located. Since capacitors consume reactive power, the installation of a capacitor bank at the load bus will have this effect. The result of adding the properly sized capacitor banks at the load side is that no reactive power would be required to flow through the lines of the system to supply the load reactance. In other words, reactive power supply by capacitor banks at the inductive load must be self-sufficient.

3. **Control of Tap-Changing and Regulating Transformers:** The flow of real and reactive power from one terminal of a transmission line toward the other terminal is dominated by the following terms:

$$P_{12} = V_1 V_2 b(\theta_1 - \theta_2)$$
$$Q_{12} = V_1 b(V_1 - V_2) \tag{2.34}$$

Expressing the same terms in per unit, we get

$$P_{12} = b(\theta_1 - \theta_2)$$
$$Q_{12} = b(V_1 - V_2) \tag{2.35}$$

These equations clearly indicate that the variables that control real and reactive power flows on a line are voltage angles and voltage magnitudes, respectively. These variables can be controlled by the use of tap changing under load (TCUL) transformers or regulating transformers. Such transformers change the number of turns used in one of the coils by changing the location of the tap on the coil of the transformer, thus changing the ratio of transformation of the transformer. This operation allows for small changes in voltage, usually ± 10%. Regulating transformers do not transform between voltage levels but rather just make small adjustments in voltage from one side to the other. With these types of transformers, voltage magnitudes and angles can be adjusted to help control the flow of real and reactive power. Tap-changing and voltage-magnitude regulating transformers are used to control bus voltages as well as reactive power flow on the transmission lines. Phase-angle regulating transformers are used to control bus voltage angles as well as real power flow on the transmission lines.

CHAPTER END PROBLEMS

2.1 For the system shown in Figure 2.15, the known system data in per unit are as follows: $Z_a = j0.5$, $Z_b = j0.25$, $Z_c = j1.2$, $Z_d = j1.5$, $Z_e = j0.25$. Find the bus admittance matrix for this system.

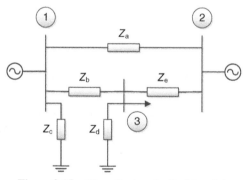

Figure 2.15   Given System for Problem 2.1.

2.2 Given the per unit voltages at bus 1 and bus 2 of the following equivalent circuit (Figure 2.16), $V_1 = 1\underline{/0°}$, $V_2 = 0.75\underline{/-9°}$. Find the complex powers S_1, S_2 at bus 1 and bus 2.

2.3 A synchronous generator is operating with $E_f = 1500\underline{/0}$ V, $V_{ts} = 1400\underline{/-80}$ V, and $X_d = j2.0\ \Omega$. (Figure 2.17). Calculate the three-phase power leaving its terminal and the load connected to its terminal in ohms.

2.4 For the given power system (Figure 2.18), $V_1 = 1.03\underline{/0°}$ (voltage at bus 1), $|V_2| = 1.02$ (voltage magnitude at bus 2), $P_2 = 0.24$ (power injection at bus 2), load at bus 3 is

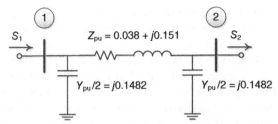

Figure 2.16 Equivalent Circuit for Problem 2.2.

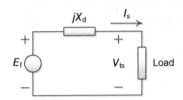

Figure 2.17 Equivalent Circuit for Problem 2.3.

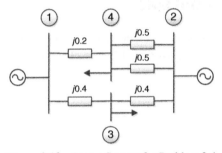

Figure 2.18 Power System for Problem 2.4.

$(0.23 + j0.19)$ p.u., and load at bus 4 is $(0.42 + j0.13)$ p.u. Using any power flow analysis software, calculate any remaining missing parameters of the system. Make any necessary assumptions for the analysis.

2.5 In the given system (Figure 2.19), a synchronous generator is connected to two loads with equal peak value of 400 A. The electric system is operating in steady-state condition

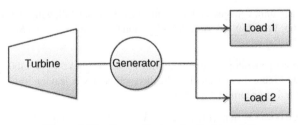

Figure 2.19 Power System for Problem 2.5.

when load 1 is suddenly disconnected from the generator. At that moment, what would happen to the turbine-generator set? What are the available control measures to prevent any negative impact on the generator as well as the entire system? Is it possible for the turbine-generator to lose load 2 as well?

2.6 Find out a major power failure (blackout) for a region or part of a country in the past. Write a short summary of this event. What did you learn?

FURTHER READING

1. Weedy BM, Cory BJ, Jenkins N, Ekanayake JB, Strbac G. *Electric Power Systems*, 5th edition. UK: John Wiley & Sons; 2012.
2. Shultz RD, Smith RA. *Introduction to Electric Power Engineering*. John Wiley & Sons; 1987.
3. McPherson G, Laramore RD. *An Introduction to Electrical Machines and Transformers*. John Wiley & Sons; 1990.
4. Toro VD. *Electric Power Systems*. Prentice-Hall; 1992.
5. Grainger JJ, Stevenson Jr. WD. *Power System Analysis*. Hightstown, NJ: McGraw Hill; 1994.
6. Saadat H. *Power System Analysis*. Hightstown, NJ: McGraw-Hill; 1999.

Chapter 3

Microeconomic Theories

The materials covered in this chapter can be found in a number of textbooks on microeconomics. In this chapter, we will cover the fundamental theories and concepts of microeconomics to the extent necessary for the reader to understand the relationship between these concepts and their applications to electric power industry in general, and to electricity markets in particular. Those readers who have a background in economics may skip this chapter. However, there will be some benefits by reviewing the materials covered here. This chapter is primarily written for those who need to have some basic understanding of economic theories that underpin the foundation, existence, and operation of electricity markets. Branches of economics that are directly applicable to electric power industry and electricity markets are microeconomics including markets and game theory, and regulatory economics.

On notation, it is customary to use P for market price and Q for quantity in economic textbooks and literature. p means the price set by a firm or paid by a consumer and q stands for a quantity of output produced by a firm or purchased by a consumer. We will follow this customary notations in this chapter so that readers will become familiar with the similar notations that are used in other references in economics. In electrical power engineering, it is also customary to use P for real power and Q for reactive power. Although notations are similar, the concepts are vastly different. We hope that the adoption of the same notations in two different fields should not create confusion over the fundamental concepts in each respective field.

3.1 PRELIMINARIES

Microeconomics is a branch of economics which studies the behavior and interaction of consumers and individual firms in a market or economy. Therefore, the study of a market or markets is the central theme of microeconomics. So, what is a market? We

Electricity Markets: Theories and Applications, First Edition. Jeremy Lin and Fernando H. Magnago.
© 2017 by The Institute of Electrical and Electronics Engineers, Inc. Published 2017 by John Wiley & Sons, Inc.

will provide some introductory remarks and key concepts of markets here. Later in the chapter, we will provide deeper theoretical treatment of markets.

Market is a collection of buyers and sellers that determine the price of a product or products by some kind of interactions such as trading or exchange, among them. In an electricity market, the buyers are consumers of electricity or representatives of such consumers, while sellers are electricity-generating resources, and the product in trade is electricity. The price of electricity in a market setting is represented by the *market-clearing price* which is typically administered by a neutral third party, such as power exchange or independent market operator which has both authority and ability to clear the market under some market-clearing rules.

There are buyers and sellers in any market. The function of a buyer is to purchase goods and services. The function of a seller is to sell such goods and services that the buyers purchase. Well functioning of a market or markets is essential to any industry. Markets can be either competitive (or perfectly competitive) or noncompetitive. This is one of the important criteria that is used to judge an electricity market whether it is well functioning. *Perfectly competitive market* can be defined as a market that has many buyers and sellers, so that no single buyer or seller can significantly affect the market price. Generally, a market that is not competitive can have adverse consequences. We will cover these concepts more in depth later in the chapter.

One of the key functions of a market is the determination of a market price. In an electricity market, the market price is the price of electricity. The determination or discovery of market price is important because buyers make their purchase decisions to consume goods and services based on the actual or expected market prices. Sellers also make similar decisions to sell goods and products based on the same prices. This important fact also holds true for electricity markets. *Market price* is a price that is determined by the balance of supply and demand in a market. In electricity markets, the balance or intersection of supply and demand for electricity determines the electricity market price. Market prices of many goods can fluctuate over time and sometimes can change quickly. This is especially true for goods or products sold in highly competitive markets. Occasionally, market prices in electricity markets exhibit similar behavior. The computation methods used to determine such electricity prices in an electricity market will be dealt in subsequent chapters.

Market definition is also another important concept in microeconomics. It defines the boundaries of a market in terms of both geographic boundary and product ranges that are necessary to be included in the market definition. In electricity markets, the geographic boundary, defined through electrical boundary and connection, as well as different electricity products will determine the market definition. For example, the electricity markets in the United States have distinct geographic boundaries, enclosed by an electrical network, which may have interconnections with other electrical networks outside the boundaries of such markets. While there may be some interregional trading across these boundaries, the market operator for each distinct electricity market will be primarily responsible for operating its own market. In terms of range of products, there are a number of electricity-related products being transacted or traded in electricity markets. Such products include energy, ancillary services (regulation service, operating reserve, etc.), and capacity. Reactive power, black start, and voltage regulation services can also be market products that have potential for trading and transaction.

Fundamental to any market mechanism is the relationship between supply and demand. This relationship is represented by supply and demand curves. The *supply curve* shows the relationship between the quantity of a good that producers are willing and able to sell and the associated price for each level of quantity. The supply curve is upward sloping. For example, the firms or suppliers are willing to produce and sell more if the price is higher. A higher price may also attract other firms to enter the market.

A typical supply curve is drawn graphically in Figure 3.1. If there is only one supplier in the market, the supply curve of that supplier will represent the total supply curve for the whole market. If there are more than one supplier, then the aggregate of supply curves of all suppliers represents the total supply curve for the entire market. Mathematically, we can represent a supply curve as a function of price as

$$Q_S = Q_S(P) \tag{3.1}$$

From the graph, we can see that a firm is willing to sell Q_1 amount of goods if the market price is P_1. Movement along the supply curve represents the response of the quantity supplied to the changes in price. A supply curve of a supplier can shift to the left or right, depending on the increase or decrease in the production cost of that supplier. The same phenomenon applies to the aggregate supply curve.

Similarly, the *demand curve* shows the relationship between quantity of a good that consumers are willing and able to purchase and the associated price for each level of quantity that they are willing to pay. The demand curve is downward sloping.

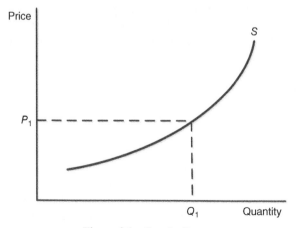

Figure 3.1 Supply Curve.

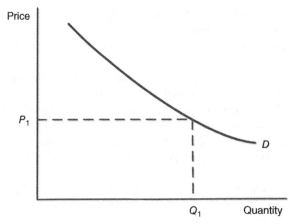

Figure 3.2   Demand Curve.

For example, consumers are willing to buy more if the price is lower. A lower price may also encourage consumers to buy and consume more quantities. It may also encourage other consumers to begin buying it.

A typical demand curve is drawn graphically in Figure 3.2. If there is only one consumer in the market, the demand curve of that consumer will represent the total demand curve for the whole market. If there are more than one consumer, which is typically the case, then the aggregate of demand curves of all consumers represents the total demand curve for the entire market. Mathematically, we can represent a demand curve as a function of price as

$$Q_D = Q_D(P) \tag{3.2}$$

In microeconomics, there are two additional types of goods and products that can affect the supply and demand curves. These two goods are *substitute* and *complementary* products. Two products are substitutes if the price increase in one product leads to an increase in quantity demanded of the other. On contrary, two products are complements if the price increase in one product leads to a decrease in quantity demanded of the other product.

Again, the market is a place where suppliers or sellers, and consumers or buyers meet and transact. A market develops when supply curve and demand curve intersect. As a corollary, a market will not develop if supply curve and demand curve never intersect. The intersection of supply and demand curves is illustrated in Figure 3.3.

The price and quantity of a product that correspond to the intersection of both supply and demand curves are known as *equilibrium price* and *equilibrium quantity*, respectively. The equilibrium price is also known as market-clearing price. The notion of supply and demand curves, and their intersection at market equilibrium is based on the assumption that the market under consideration is relatively competitive. In

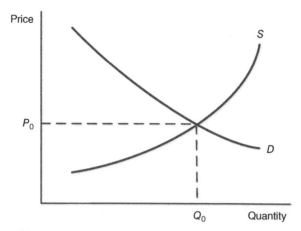

Figure 3.3   Intersection of Supply and Demand Curves.

most markets, both supply and demand curves can shift over time for various reasons. With new supply and demand curves, new equilibrium price and quantity will appear. Therefore, equilibrium price and equilibrium quantity exhibit dynamic behavior.

Related to the concepts of supply and demand curves, the concept of supply and demand elasticities is also important. The supply and demand elasticities show the sensitivity of either supply or demand to a change in another variable. For example, the consumers will change their demand quantity if the price changes. Percentage change in quantity demanded of a good when the price increases by one percent is known as *price elasticity of demand*. Price elasticity of demand is typically negative, meaning if the price of a good increases, the quantity demanded falls. Price elasticity of demand should be measured at a particular point on the demand curve and will have different values at different points on the demand curve. There are some products that show infinitely elastic demand or completely inelastic demand. Similarly, the *price elasticity of supply* is the percentage change in quantity supplied when the price is increased by one percent. Price elasticity of supply is usually positive because producers produce more output when price increases. Price elasticities of both supply and demand can change over time. The price elasticity of demand or supply can be written as

$$\epsilon_p = \frac{\Delta Q/Q}{\Delta P/P} \qquad (3.3)$$

3.2 THEORY OF CONSUMER BEHAVIOR

Theory of consumer behavior explains how a consumer with limited income decides which goods and services to buy based on his or her preference. The key to this theory lies in the concept of utility function. The *utility function* describes the level of satisfaction or happiness one can obtain from consuming certain products. While it is difficult to objectively measure a person's satisfaction or level of well being from consumption of a product, the concept of utility function is useful and fundamental to understand the consumer demand function.

Another key concept related to utility function is that a consumer satisfaction (utility) is maximized when the marginal benefit is equal to the marginal cost. *Marginal benefit* is the benefit received by the consumption of one additional (marginal) unit of a good. In an electricity market, the marginal benefit is the benefit that a consumer would receive by the consumption of one additional unit of electricity. In a similar fashion, *marginal utility* measures the additional satisfaction obtained from consuming one additional unit of a good. *Marginal cost* is the cost to produce (for suppliers) or purchase (for consumers) one additional unit of a good. In an electricity market, the marginal cost means the cost to produce or purchase one additional unit (in kW or MW) of electricity.

Even though consumers will enjoy increased utility as they consume more, there is a limit to this aspect. In other words, the value of utility can decline as more of certain goods are consumed. This is because the additional goods will make less and less utility for the consumers. This concept is known as *diminishing marginal utility*. More formally, the law of diminishing marginal utility states that as a person increases consumption of a product—while keeping consumption of other products constant—there is a decline in the marginal utility that person derives from consuming each additional unit of that product. For example, for a person who does not own a car yet, having the first car will add significant amount of utility (satisfaction). However, assume that the person owns more than one car. Possession of each additional car will not add as much satisfaction as owning the first or second car. In fact, owning more than one car can add significant expenses (burden) that the utility can become negative.

3.2.1 Individual and Market Demand

Theory of demand is based on the assumption that consumers maximize utility by consuming goods and products. The theory explains the behavior of demand for a particular product. For example, the demand for a good depends on its price, prices of other goods, and income of the consumers.

In general, consumers consume more of a good if the price of the good decreases. It is important to understand the nature of individual demand curve. Individual demand curve shows the relationship between the quantity of a good that a single consumer will buy and the price of that good. There are two important properties associated with the individual demand curve. First, the level of utility changes as we move along the curve. The lower the price of a product, the higher the level of utility by consuming more of it. On the other hand, as the price of a product increases, the consumer will demand less, thus, the utility for the consumer decreases. Second, at every point on

Price

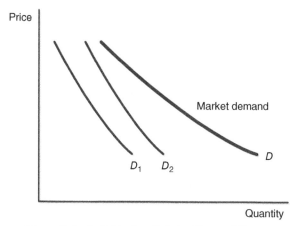

Market demand

D_1 D_2

D

Quantity

Figure 3.4   Individual and Market Demand Curves.

the demand curve, the consumer is maximizing its utility by satisfying the condition that marginal rate of substitution of one product for another equals the ratio of prices of both products.

There are two kinds of goods generally defined in a market. They are normal goods and inferior goods. A good is known as a *normal good* if consumers want to buy more of it as their income increases. Hence, the income elasticity of demand for a normal good is positive. For *inferior goods*, consumers want less of them as their income increases. The income elasticity of demand for this kind of good is negative.

Market demand shows how much of a good that consumers overall are willing to buy as its price changes. *Market demand curve* shows a relationship between the quantity of a good that all consumers in a market will buy and its price. It can be derived as a sum of the individual demand curves of all consumers in a particular market. Market demand curve is the horizontal summation of the demands of each consumer. It is a downward sloping curve. This is illustrated in Figure 3.4. D_1 represents the demand curve for consumer 1, and D_2 represents the demand curve for consumer 2. The summation of demand curves of both consumers is the market demand curve which is represented by D. There are two important points to note about market demand curve. First, the market demand curve will shift to the right as more consumers enter the market and consume more of the same products. Second, the factors that influence the demands of many customers will also affect the market demand curve.

Previously, we have introduced the concepts of demand curves and price elasticity of demand. Demand is said to be *inelastic* if the absolute value of price elasticity of demand is less than one. In contrast, the demand is *elastic* if the absolute value of price elasticity of demand is greater than one. *Isoelastic* demand means that the price elasticity of demand is constant all along the demand curve.

3.2.2 Consumer Surplus

Consumer surplus of an individual consumer is the difference between the maximum amount that the consumer is willing to pay for a good (known as *reservation price*) and the amount that the consumer actually pays. In other words, consumer surplus is the total benefit from consumption of a product, less the total cost of purchasing it. Hence, aggregate consumer surplus is the measure of the summation of consumer surplus of all consumers who buy a particular good.

Consumer surplus can be calculated from the demand curve. Recall that the demand curve shows the relationship between the willingness of consumers to pay for quantity of a product at different prices. So, aggregate consumer surplus in a market is the area between the market demand curve and above the price line. This is illustrated in Figure 3.5.

The consumer's surplus is a classic tool to measure changes in social welfare. If $x(p)$ is the demand for some good as a function of its price, the consumer's surplus associated with a price movement from p^0 to p^1 is

$$CS = \int_{p^0}^{p^1} x(t)dt. \tag{3.4}$$

This is the area to the left of the demand curve between p^0 and p^1. Depending on the representation of consumer's preference by different utility functions, consumer's surplus may be either a reasonable approximation or exact measures of welfare change.

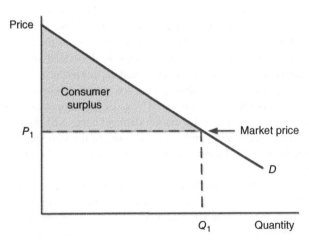

Figure 3.5 Consumer Surplus.

3.3 THEORY OF A FIRM

Theory of a firm describes how a firm makes production decisions that minimize its cost so as to maximize its profit. The fundamental problem faced by a firm is how much of output it should produce. Before answering this question, let us analyze the objective of a firm. It is reasonable to assume that the goal of a firm is to maximize its profit.

3.3.1 Profit Maximization Assumption

Even though it is not always the case, it is reasonable to assume that the goal of each firm in a market is to maximize its profit. We can write this objective as

$$\max \quad \pi(q) = R(q) - c(q), \tag{3.5}$$

where $R(q)$ is the revenue as a function of quantity q, $c(q)$ is the cost function, and $\pi(q)$ is the profit function. To maximize its profit, a firm selects output level for which the difference between revenue and cost is the greatest. The first-order optimality condition for profit maximizing requires that

$$\frac{\partial \pi(q)}{\partial q} = R'(q) - c'(q) = 0$$
$$R'(q) = c'(q), \tag{3.6}$$

where $R'(q)$ is marginal revenue and $c'(q)$ is marginal cost. Thus, the profit is maximized for a firm when its marginal revenue equals to its marginal cost. This is one of the most important concepts in the theory of a firm.

3.3.2 Cost of Production

Various definitions of *cost* are in order to understand the advanced topics on the market later in the chapter. Economists are generally concerned with *economic cost* which is the cost of utilizing resources in production, including opportunity cost. *Opportunity cost* is the cost associated with opportunities that are foregone by not putting the firm's resources to the best alternative use. When considering economic cost, we should not include the *sunk cost* which is the expenditure that has been made and cannot be recovered.

The *total cost* (or total economic cost) of production has two components: *fixed cost* and *variable cost*. Fixed cost is the cost that does not change with the level of output. For example, expenditures on plant buildings do not change with the amount of output produced for a manufacturing firm. Variable cost is the cost that varies as output varies. Wages and salaries of employees, for example, are dependent on the level of output, hence they are variable costs.

In addition, it is also important to understand the concepts of marginal cost and average cost. *Marginal cost* measures the increase in cost, resulting from the production of one additional unit of output. Marginal cost tells us how much it will cost to expand output by one more unit. It is also known as *incremental cost*. *Average total cost* or simply *average cost* is the firm's total cost divided by its level of output. It tells us the per-unit cost of production. Average cost has two components: *average fixed cost* and *average variable cost*. Costs are also different in the short run versus long run. For example, costs that are fixed in the short run may not be fixed over a longer time horizon. In fact, all input costs can be variable in the long run.

Firms also enjoy *economies of scale* when spending two times the input costs more than doubles its output. On the other hand, firms would face *diseconomies of scale* when spending two times the input costs less than doubles its output.

In general, firms use various combinations of input to produce output given the production technology and input cost. Input–output relation for a firm is shown in Figure 3.6a. From this relationship, marginal product and average product curves of that product are also shown in Figure 3.6b.

From the figure, we can see that marginal cost is positive as long as output is increasing, but becomes negative when the output is decreasing. Marginal cost crosses the horizontal axis of the graph at the point of maximum total production. Marginal cost becomes negative when output starts to fall as more input is needed. Average cost is increasing when marginal cost is greater than average cost. Average cost is decreasing when marginal cost is less than average cost. Marginal cost and average cost equal when average cost is at maximum.

Firms choose optimal combination of inputs while minimizing cost. A firm's costs depend on its rate of output, as well as the price of inputs. A firm's costs change over time.

So far, we have known that it costs firms to produce output in terms of input cost. The relationship that shows how much it costs for a firm to produce certain level of output represents the cost function. Mathematically, a cost function (cost of production) can be linear, quadratic or cubic form. From these cost functions, we can derive associated marginal cost functions. For example, the linear cost function can be written as

$$c(q) = \beta q \qquad (3.7)$$

where β is constant. Similarly, the quadratic cost function can be expressed as

$$c(q) = \beta q + \gamma q^2, \qquad (3.8)$$

where both β and γ are constants. The marginal cost functions for these cost functions can also be shown as

$$c' = \beta \quad \text{(for linear cost function)}$$
$$c' = \beta + 2\gamma q \quad \text{(for quadratic cost function)} \qquad (3.9)$$

In general, it is difficult to measure a cost function for various reasons.

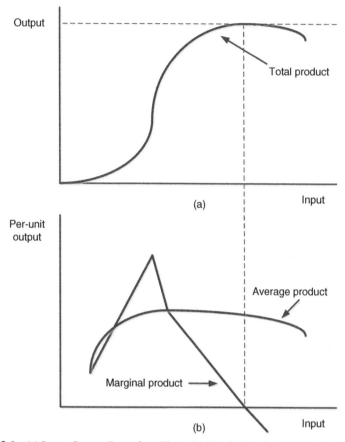

Figure 3.6 (a) Input–Output Curve for a Firm; (b) Marginal and Average Product Curves.

3.3.3 Demand and Marginal Revenue for a Competitive Firm

In a competitive market, each firm is a price taker for the reasons that will be described in Section 3.4. It is important to distinguish between market demand curves and the demand curves faced by individual firms. Because the firm is a price taker, the demand curve d facing an individual competitive firm is given by a horizontal line. The firm will decide how much level of output q to produce for a given price.

Figure 3.7b shows how much of a product that all consumers will buy at each possible price. It is downward sloping because consumers buy more product at a lower price. Figure 3.7a shows that a demand facing a firm is horizontal because a firm's sales will not affect the market price. Again, market price is determined by the interactions of all firms and consumers in the market, not by the output decision of a single firm. When an individual firm faces a horizontal demand curve, it can sell an

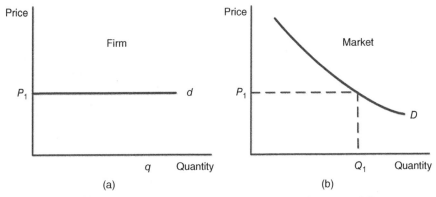

Figure 3.7   (a) A Demand Curve Facing a Firm; (b) Market Demand Curve.

additional unit of output without the need to lower the price. So, when a firm sells an additional unit of output, the firm's total revenue will increase with the additional revenue created by this output. Hence, the firm's marginal revenue is constant at market price. Average revenue of the firm will be the same as marginal revenue.

This is an important fact for electricity markets. For an electricity market to be competitive, the size of the market should be sufficiently large so that the output decision of any firm will have little or no influence on the market price. The ability to influence market price is also a key factor in determining whether a firm has market power. We will treat the topic of market power more extensively in a later chapter.

Hence, the demand curve facing an individual firm in a competitive market is both its average revenue curve and marginal revenue curve. Along this demand curve, marginal revenue, average revenue, and market price are all the same.

$$R'(q) = \text{market price} \qquad (3.10)$$

Earlier, we learned that the profit of a firm is maximized when its marginal revenue is equal to its marginal cost. Here, we note that marginal revenue is equal to the market price along the demand curve facing a firm. As a result, in a perfectly competitive market, a firm would choose an output where its marginal cost equals to the market price.

$$c'(q) = R'(q) = \text{market price} \qquad (3.11)$$

In this case, competitive firms take the market price as fixed.

3.3.4 Choosing Output in the Short Run

In the short run, capital is fixed but variable outputs can be changed to support profit-maximizing output. The graph for profit-maximizing output is shown in Figure 3.8.

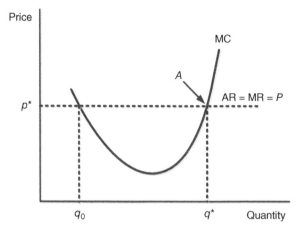

Figure 3.8 Profit-Maximizing for a Firm in the Short Run.

In the figure, profit is maximized at point A, where a demand curve facing the firm (market price) intersects with the firm's marginal cost curve. This is the point where marginal revenue equals marginal cost. q^* represents the profit-maximizing output for the firm. At q_0, marginal benefit also equals marginal cost. However, the firm can increase profit by producing beyond q_0 output because marginal cost is below marginal revenue. So, profit-maximizing condition can be stated more precisely—the profit is maximized when marginal revenue equals marginal cost at a point at which marginal cost curve is rising. This profit-maximizing output decision for a firm holds whether the market is perfectly competitive or not. Hence, the concept of marginal cost of production is important for firm's economic decision.

Short-run market supply curve shows the amount of output that the industry will produce in the short run for every possible price. Industry output is the sum of all the quantities supplied by all of its individual firms. Therefore, market supply curve can be obtained by adding supply curves of each firm. This is illustrated in Figure 3.9.

3.3.5 Producer Surplus

Producer surplus of a firm is the difference between the market price and marginal cost of producing the units. As shown in Figure 3.10, the producer surplus is given by the area under the firm's horizontal demand curve and above its marginal cost curve from zero output to the output at which profit is maximized. Alternately, producer surplus can be defined as the difference between the firm's revenue and its total variable cost. Similar to consumer surplus of a market, the producer surplus of a market can be calculated from the market supply curve. Recall that supply curve shows the relationship between the marginal cost of producers to supply for quantity of a product at different prices. So, aggregate producer surplus can be obtained by adding producer surpluses of all firms.

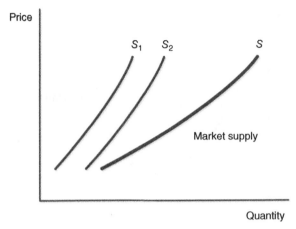

Figure 3.9   Individual Firm and Market Supply Curves.

3.3.6 Social Welfare

Social welfare is the sum of consumer surplus and producer surplus for a market or industry. It is a measure of welfare benefit of a competitive market as well. Economists are primarily interested in welfare effects of any change in policy or market rules. Welfare effects include gain or loss to consumers and/or producers. Economic efficiency in a market means the maximization of aggregate consumer surplus and producer surplus for that market. Achieving economic efficiency in a competitive market is the goal to be attained. Any net loss of both consumer and

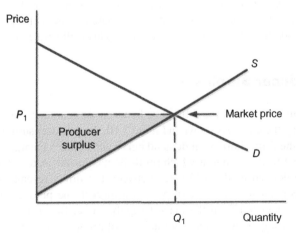

Figure 3.10   Producer Surplus.

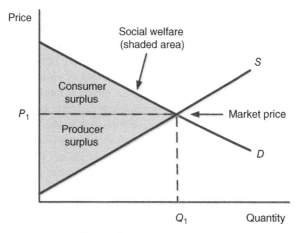

Figure 3.11 Social Welfare.

producer surpluses is known as *deadweight loss*. Economic efficiency will be reduced by the amount equal to that deadweight loss.

Assume that a market has only one consumer and one firm. Let $CS(q) = u(q) - pq$ be the consumer surplus associated with the given level of output. Also, let $PS(q) = pq - c(q)$ be the producer surplus earned by that firm. Then, the maximization of total surplus (social welfare) is

$$\max \quad CS(q) + PS(q) = [u(q) - pq] + [pq - c(q)] \tag{3.12}$$

or equivalently as

$$\max \quad u(q) - c(q) \tag{3.13}$$

This is equivalent to saying that the social welfare is maximized by the competitive equilibrium of output in a market. The social welfare for a market is graphically shown in Figure 3.11.

3.4 THEORY OF A MARKET

In previous sections, we briefly introduced the characteristics of a market. We learned that in any market, firms make decisions on price and output level to maximize the profits. In many markets, firms who compete with each other have at least some level of market power. These firms have control over price and can profitably charge a price that exceeds their marginal costs. In this section, we will provide more detailed analysis of different types of market structures and nature of competition among firms within these markets.

Previously, we have explained the behavior of consumers and firms, showing their optimal decisions when market prices are fixed and beyond their control. In this section, we will explore the outcome of these behaviors when consumers and firms meet in different market settings. We will consider the determination of equilibrium price and quantity in a single market or a group of closely related markets by the actions of agents (consumers and firms) for different market structures. Such equilibrium analysis is known as *partial equilibrium analysis* because the analysis focuses on a single market or a group of closely related markets, with the implicit assumption that changes in the markets under consideration will not affect the prices of other goods and hence upset the equilibrium that holds in other markets. Treatment of all markets simultaneously is generally covered in the general equilibrium theory.

The focus of this section is on modeling the market behavior of the firm. We will explore the question on price determination by firms when they sell their products or goods or by consumers on how much they are willing to pay. In certain situations, price-taking behavior would be a good approximation to an optimal behavior, while in other situations, price-setting behavior may be the model. Let us start with an idealistic case of *perfect competition*. Then, we will study markets in which agents have some level of market power. Such market structures include *pure monopoly*, *monopolistic competition*, *oligopoly*, and *monopsony*.

We have stated that the determination of price in a market setting is one of the most important objectives of a market. Market prices perform three functions in organizing economic activities in a free market setting: (1) transmission of information about production and consumption, (2) provision of right incentives, and (3) determination of distribution of income and products. We will start with perfectly competitive markets or perfect competition.

3.4.1 Perfect Competition

Perfect competition is an idealistic market structure. Although not very realistic, a model of perfect competition is useful for studying a variety of markets. There are four basic assumptions in perfectly competitive markets: (1) price taking, (2) product homogeneity, (3) free entry and exit, and (4) symmetric information. These assumptions are further explained below.

1. **Price Taking**: When each individual firm in a market sells a sufficiently small amount of total market output, the impact on the market price by the firm's action is almost zero. Thus, each firm is a *price taker* in this market. So, the axiom is that firms in a perfectly competitive market are price takers. This assumption applies to consumers as well.

2. **Product Homogeneity**: The assumption of price taking depends on another assumption, which is the homogeneity of products. Products in a market are homogeneous when firms produce the identical or nearly identical products. In other words, when the products of all firms in a market are perfectly substitutable with one another, no firm can raise the price of its products above the price of other firms without losing most or all of its business. Homogeneous products are also known as *commodities*.

   In contrast, when products are heterogeneous, each firm has an opportunity to raise its price above that of its competitors without losing all of its sales. The assumption of product homogeneity ensures that there is a single market price, consistent with supply–demand analysis.

3. **Free Entry and Exit**: Entering or exiting a market is free if there are no special costs that make it difficult for a new firm to enter or for an existing firm to exit an industry or a market. As a consequence, buyers can easily switch from one producer to another, and suppliers can easily enter or exit a market.

4. **Symmetric Information**: It is important for all players (firms and consumers) in a market to have all information necessary to make correct economic decisions. If one firm or one consumer has more information than others, markets cannot be perfectly competitive. Symmetric information means that consumers have all readily available information about prices and products from competing suppliers and can access this information at zero cost. In other words, the transaction costs associated with searching for the required information about prices are very low. Likewise sellers have perfect knowledge about their competitors.

If at least one of these assumptions does not hold, the market under consideration cannot be perceived as perfectly competitive. If a market is perfectly competitive, each firm in that market faces a perfectly horizontal demand curve for a homogeneous product in an industry that it can freely enter or exit. In a highly competitive market, firms face highly elastic demand curves and relatively easy entry and exit. In reality, it is not an easy matter to determine whether a market is close to being perfectly competitive. The fact that there are many firms in an industry or a market is not sufficient to affirm that a market/industry is perfectly competitive. In contrast, the presence of a few firms in a market does not preclude that the market is competitive. Demand elasticity of a product in a market will make the demand curve facing each firm to be nearly horizontal. Therefore, there is no simple way to determine whether a market is highly, if not perfectly, competitive.

3.4.1.1 A Competitive Firm

A competitive firm is a firm that takes the market price of a product it sells as given. Let $\bar{p}$ be the market price. Then, the demand curve facing a competitive firm will be

$$D(p) = \begin{cases} 0 & \text{if} \quad p > \bar{p} \\ \text{any amount} & \text{if} \quad p = \bar{p} \\ \infty & \text{if} \quad p < \bar{p} \end{cases} \tag{3.14}$$

In a competitive market, each firm has to sell its product at the market price. If a firm attempts to sell its product at a price which is above the market price, it would lose all of its customers. On the other hand, if a firm sets its price below the market price, it will lose some revenues from the sale (not profit maximizing). Therefore, each firm must take the price as given, and treat it as an exogenous variable when it is making its supply decision.

If a competitive firm takes the market price as given, its profit maximization problem is to choose an output level q^* by solving

$$\max \quad \pi = pq - c(q), \tag{3.15}$$

where q is firm's output, p is the market price of the product, and $c(q)$ is the cost function of producing this product. The first-order condition for optimality requires

$$p = c'(q) \equiv MC(q) \tag{3.16}$$

The first-order optimality condition becomes a sufficient condition if the second-order condition is satisfied:

$$c''(q) > 0 \tag{3.17}$$

Thus, the supply function of a competitive firm is determined by satisfying both first-order and second-order conditions. At any price p, the firm will supply an amount of output $q(p)$ such that $p = c'(q(p))$. By $p = c'(q(p))$, we have

$$1 = c''(q(p))q'(p) \tag{3.18}$$

and hence,

$$q'(p) > 0 \tag{3.19}$$

which upholds the law of supply. The condition $p = c'(q^*)$ is the first-order condition that characterizes the optimum only if $q^* > 0$, that is, only if q^* is interior optimum. In the short run,

$$c(q) = c_v(q) + F \tag{3.20}$$

where $c_v(q)$ is the variable component of the production cost function and F is the fixed portion of the same production cost function. The firm should produce if

$$pq(p) - c_v(q) - F \geqq -F \tag{3.21}$$

and we have

$$p \geq \frac{c_v(q(p))}{q(p)} \equiv \text{AVC} \tag{3.22}$$

Equation (3.22) means that the necessary condition for a firm to produce a positive amount of output is that the price must be greater than or equal to the average variable cost of the firm.

In general, the supply curve for the competitive firm is given by $p = c'(q)$ if $p \geqq [c_v(q(p))/q(p)]$ and $q = 0$ if $p \leqq [c_v(q(p))/q(p)]$. The supply curve is represented by upward sloping portion of the firm's marginal cost curve if the market price is greater than or equal to the firm's average variable cost. However, the supply curve will be zero if the market price is less than the firm's average variable cost.

Assume that there are j firms in a market. The aggregate supply function for the market is the summation of supply functions of all firms. This can be written as

$$\hat{q}(p) = \sum_{j=1}^{J} q_j(p), \tag{3.23}$$

where $q_j(p)$ is the supply function of firm j for $j = 1, \ldots, J$. The industry supply function measures the relationship between industry output and the associated cost of producing this output.

The aggregate demand function for the market is also the summation of demand functions of all consumers. This can be written as

$$\hat{x}(p) = \sum_{i=1}^{I} x_i(p), \tag{3.24}$$

where $x_i(p)$ is the demand function of consumer i for $i = 1, \ldots, I$.

Market price is determined when the total amount of output that firms wish to supply equals to the total amount of output that consumers wish to consume. Equilibrium price p^* is the solution where aggregate quantity demanded equals the aggregate quantity supplied, which can be written as

$$\sum_{i=1}^{I} x_i(p) = \sum_{j=1}^{J} q_j(p) \tag{3.25}$$

3.4.2 Pure Monopoly

The market structure at the opposite end of perfect competition is pure monopoly. In a pure monopoly, there is only one seller for a specific product. This single seller is known as monopolistic firm. As in the case of firms in a perfectly competitive market, a monopolistic firm therefore has to decide how much to produce and at what price to sell them. However, the amount of output that the firm can sell depends on the price it sets. This relationship between the demand and price for a monopolistic firm can be represented by market demand function for output $q(p)$ (q here because the market demand curve is the same as firm's demand). This market demand function will give us information about how much output consumers will demand based on the price that a monopolistic firm charges. The same relationship can be represented by inverse demand function $p(q)$, which shows the price that consumers are willing to pay for q amount of output. Here, we can write the monopolistic firm's objective of maximizing profit which is the difference between the revenue and cost which are functions of output as

$$\max \quad \pi = R(q) - c(q) = p(q)q - c(q) \tag{3.26}$$

The first-order optimality condition for profit maximization requires that marginal revenue must equal marginal cost:

$$p(q^*) + p'(q^*)q^* = c'(q^*) \tag{3.27}$$

The intuition behind this condition is that if the monopolistic firm decides to produce one additional unit of output, the revenue will increase by $p(q^*)$. However, this increased output will also decrease the price by $p'(q^*)$ which will be lost on all output sold. The combination of these two effects will determine the marginal revenue of the firm. The monopolistic firm will continue to produce more output until the marginal revenue equals the marginal cost.

The first-order optimality condition for profit maximization can also be expressed by using the price elasticity of demand. Recall that the price elasticity of demand can be written as

$$\epsilon(q) = \frac{p}{q(p)} \frac{dq(p)}{dp} \tag{3.28}$$

Price elasticity of demand is always negative because $dq(p)/dp$ is negative. Therefore, the equality condition of marginal revenue and marginal cost can be written as

$$p(q^*)\left[1 + \frac{q^*}{p(q^*)} \frac{dp(q^*)}{dq}\right] = p(q^*)\left[1 + \frac{1}{\epsilon(q^*)}\right] = c'(q^*) \tag{3.29}$$

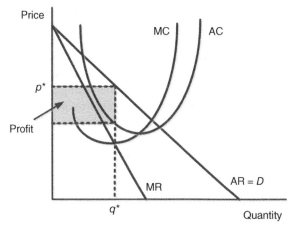

Figure 3.12   Profit Maximization by a Monopolistic Firm.

Equation (3.29) shows that a monopolistic firm would charge a price over marginal cost by an amount equal to the markup, which is a function of price elasticity of demand. This profit maximization scheme by a monopolistic firm can be graphically illustrated in Figure 3.12.

Assume that the inverse demand curve is linear, given by $p(q) = a - bq$. Thus, the revenue function becomes $R(q) = aq - bq^2$, and the marginal revenue function can be described as $R'(q) = a - 2bq$. The marginal revenue curve has the same intercept on the vertical axis as the demand curve, but is twice steeper than the demand curve. The figure shows demand curve, marginal revenue, marginal cost, and average cost curves. The optimal output q^* for a monopolistic firm is the point of intersection between marginal revenue and marginal cost. The price $p(q^*)$ (or p^* in the figure) that the monopolistic firm will charge is the point of intersection between the demand curve and the vertical line of optimal output. Therefore, the optimal revenue is the product of optimal price and optimal output, given by $p(q^*)q^*$. The cost of producing optimal output is the product of optimal output amount and average cost of production for that level of output. The shaded area in the graph is the optimal profit for the monopolistic firm.

In welfare economics, a market condition is known as *Pareto efficient* if one cannot be made better off without making someone else worse off. Therefore, a monopolistic firm is not Pareto efficient because a monopolistic firm can make itself better off without making the customers worse off.

3.4.3 Monopolistic Competition

In a monopolistic competition, there are many firms and entry by new firms is not restricted. These conditions are similar to perfectly competitive markets. Since there are many sellers, each seller would reasonably assume that its actions will not materially affect others. However, product is differentiated in markets with monopolistic competition. In other words, each firm produces its own differentiated product. The extent of this product differentiation will determine the amount of monopoly power that the firm would enjoy. Because of this monopoly power, a seller in a monopolistic competition can raise the price of its products without losing all of its customers.

Assume that there are more than one monopolistic firm in a market. In this case, other monopolistic firms will respond to the price or quantities set by one monopolistic firm. Hence, the monopolistic firm can no longer assume that demand curve for its product depends only on the price it sets. Assume also that there are n monopolistic firms who sell similar products. The price that consumers are willing to pay for the output of the firm i depends on the output of firm i and output of other firms. This inverse demand function can be written as $p_i(y_i, y)$, where $y = (y_1, \ldots, y_n)$. The profit maximization problem for firm i can be formulated as

$$\max \quad \pi_i = p_i(q_i, q)q_i - c(q_i) \tag{3.30}$$

Assuming that other firms behavior is constant, each firm i will choose its level of output q_i^* so that it will satisfy the condition:

$$p_i(q_i^*, q) + \frac{\partial p_i(q_i^*, q)}{\partial q_i}q_i^* - c_i'(q_i^*) \leq 0 \quad \text{(equality if} \quad q_i^* > 0) \tag{3.31}$$

For each combination of output for the firms $q_1, \ldots, q_n$, there must be some optimal output for firm i. Assume that this optimal choice of output is noted by $Q_i(q_1, \ldots, q_n)$. For the market to reach equilibrium, each firm's prediction about the output level of other firms must be consistent with the output that other firms produce. Therefore, for $(q_1^*, \ldots, q_n^*)$ to be in equilibrium, the following must be satisfied:

$$q_1^* = Q_1(q_1^*, \ldots, q_n^*)$$
$$\vdots$$
$$q_n^* = Q_n(q_1^*, \ldots, q_n^*) \tag{3.32}$$

That means q_1^* must be optimal choice for firm 1 if it assumes that the other firms are going to produce $(q_2^*, \ldots, q_n^*)$. This must be true for all other firms. Therefore, the equilibrium condition $(q_1^*, \ldots, q_n^*)$ in a monopolistic competition must satisfy

$$p_i(q_i^*, q) + \frac{\partial p_i(q_i^*, q^*)}{\partial q_i}q_i^* - c_i'(q_i^*) \leq 0 \quad \text{(equality if} \quad q_i^* > 0) \tag{3.33}$$

At this optimal point, the marginal revenue equals marginal cost for each firm, given the output of all other firms. In this short-run equilibrium, firm i is making positive profits. Therefore, other firms will enter the market, thus reducing the profits of all other firms. Therefore, in the long run, firms would enter the market until the profits of all firms were reduced to zero. This long-run competitive equilibrium is similar to the situation in perfectly competitive markets.

3.4.4 Monopsony

In a perfectly competitive market, no firms have any influence over the market price. Therefore, firms have to take the market price as given. Market price is also equal to the marginal cost (supply curve) of all firms combined in a competitive market environment. However, if a firm has some level of market power, the firm can charge a price that is above its marginal cost. This type of market power, known as monopoly power, represents an ability of a firm to set price above marginal cost. We will see later that firms in Cournot or Stackelberg model have some level of monopoly power so that these firms can charge a price that is well above competitive levels.

There is also a market power associated with buyer side of the market. The market power associated with the buyer side of the market is known as *monopsony power*. Monopsony refers to a market dominated by a single buyer. A monopsonist has the market power to set the price of whatever it is buying. Monopsony power allows the buyer to purchase a good at a price that is generally less than that would otherwise prevail in a competitive market. The behavior of a monopsonist is similar to that of a monopolist. Another related term is *oligopsony* which refers to a market with only a few buyers. This is the counterpart of oligopoly.

Recall that individual demand curve determines the marginal value (marginal utility) of a consumer. Therefore, the marginal value schedule of an individual buyer is the individual demand curve for the good. Marginal principle stated that individual utility is maximized when the marginal benefit equals marginal cost. Marginal cost is the additional cost of buying one more unit of a good. However, the value of marginal cost depends on whether the buyer is a competitive buyer or a buyer with monopsony power.

For example, a buyer in a competitive market has no influence over the price of a good. Therefore, the price of a good does not depend on how many units of good that a single buyer purchases. The competitive price represents the market price of the good. Hence, the marginal cost for any individual buyer is the same as market price of a good prevailing in the competitive market. If there is only one buyer, the market supply curve is not the marginal cost curve. The market supply curve shows how much a buyer must pay per unit, as a function of the total number of units that are purchased. In other words, the market supply curve is the average cost curve.

Assume that a firm is a single buyer of some input good. Let $w(x)$ be the inverse supply curve of producing that input factor x. Also let $u(x)$ be the utility function for that firm (buyer). The classic profit maximization problem becomes

$$\max \quad \pi(x) = u(x) - w(x)x \tag{3.34}$$

Then, the first-order optimality condition for this problem will be

$$\pi'(x) = u'(x^*) - w'(x^*)x^* - w(x^*) = 0$$
$$u'(x^*) = w'(x^*)x^* + w(x^*) \tag{3.35}$$

Thus, the profit maximization for that purchasing firm implies that marginal utility $u'(x^*)$ of purchasing this input factor must be equal to its marginal cost $w'(x^*)x^* + w(x^*)$. One can also see that if the price paid by the purchasing firm is too low, the low price would make the business of some suppliers uneconomic, and hence there would be some unserved demand.

Another way to write this optimality condition is

$$u'(x^*) = w(x^*)[1 + 1/\epsilon], \tag{3.36}$$

where ϵ is the price elasticity of supply. If ϵ goes to infinity, the behavior of a monopsonist will look like that of a pure competitor.

Similar to monopoly power, sources of monopsony power include elasticity of market supply, number of buyers in the market and interaction among buyers. Again, as with any market power, there is no simple way to predict how much monopsony power buyers will have in a market. Buyers concentration is one indicator to determine monopsony power. Buyers concentration shows the percentage of sales that go to different buyers, typically to largest buyers. For example, the market in which only four or five buyers account for all or nearly all of the sales would be quite concentrated. In this market, there could be considerable amount of monopsony power.

With markets, market failure is a real possibility. For instance, in some markets, markets do not produce market prices that reflect the true intersection of supply and demand curves. In this case, market fails. In other words, there is no market price that can provide proper signals to both consumers and producers, creating an inefficient market. As a consequence, the aggregate consumer and producer surpluses are not maximized. Two instances in which markets can fail are externalities and lack of information.

Most of the markets in real life fall into a market structure, known as *oligopoly*. In this type of market, the interactions and responses by firms against other firms are important for market outcome. It would be useful to introduce the concept of game theory before we expand on the oligopolistic market structures.

3.5 GAME THEORY

A game is any situation in which players, for example, firms in a market, make strategic decisions, that is, decisions that take into account of each other's actions and responses. Examples of games are (1) firms competing with each other by setting prices and (2) a group of suppliers bidding against each other at an auction for electricity. *Game theory* is the study on the behavior of strategic interactions among economic agents who make economic decisions in a more complex environment. Most economic behavior can be viewed as special cases of game theory. Strategic decisions result in *payoffs* to the players. Payoffs are outcomes that generate rewards or benefits. For price-setting firms, the payoffs are profits. For bidders at an auction, the payoffs for the winner is her consumer surplus.

Application of game theory has been an important development in microeconomics. Game theory is used to understand how markets evolve and operate. Game theory is also used in auction design and bidding strategies. Much progress has been made in the field of game theory which has become a necessary tool for economic analysis.

The key objective of game theory is to determine the optimal strategy for each player. A *strategy* is a rule or plan of action for playing the game. The optimal strategy for a player is the one that maximizes his/her expected payoff. The key assumption in game theory is that the players are *rational*, that is, players act to improve or maximize their own profits or benefits.

The economic games that firms play can be either *cooperative* or *noncooperative games*. In a cooperative game, players can negotiate binding contracts that allow them to plan joint strategies and decisions. In a noncooperative game, such negotiation and binding contracts among firms are not possible. In some markets, cooperative behavior can prevail. However, in other markets, noncooperative behavior may be predominant. Cooperations are more difficult in a market in which there are many competing firms. Sometimes, it is difficult for firms to cooperate because demand and cost conditions change quickly. Auctions and electricity markets are examples of noncooperative games.

So, how can players decide on the best strategy in playing a game? While there are many examples of two players' games (duopoly) in the literature of game theory, the game of *Prisoner's Dilemma* is one of the most classic. In this game, there are two prisoners who are held in separate prison cells for allegedly committing a crime. Each is told that if he confesses and the other prisoner does not, he will get lighter sentence of just 1-year in prison. However, if he does not confess and the other confesses, he will get 10-year sentence. If both confess, they will get 5-year each. If neither confesses, it is still possible to convict both of them to 2-year sentences. Each player wishes to minimize the time he would spend in jail. The payoff (negative means time in jail) matrix is shown in Figure 3.13.

Based on the payoff matrix, each prisoner has an incentive to confess regardless of what he believes the other prisoner will do, because confessing will give each prisoner a lighter sentence compared with the choice of not confessing. So, the choice of confession turns out to be the best strategy for each prisoner. Since confession is the best strategy for both prisoners, both will confess. In the end, both will receive 5 years of sentence. In this case, confession is the *dominant strategy* for both prisoners, because this strategy gives an optimal outcome for each prisoner regardless of the choice made by other prisoner. A firm has no dominant strategy when its optimal

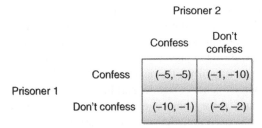

Figure 3.13   The Prisoner's Dilemma.

decision depends on what other firms would do. In general, dominant strategies are stable. This outcome (confess, confess) is also a *dominant strategy equilibrium* because each prisoner will predominantly choose this strategy (confess). In fact, a dominant strategy equilibrium is a special case of *Nash equilibrium*. However, not all Nash equilibria are dominant strategy equilibria. A dominant strategy equilibrium is a necessary solution to the game because each player has a unique optimal choice.

If a player consistently chooses only one specific strategy with probability one throughout the game, this strategy is called *pure strategy*. On the other hand, if a player makes random choices among two or more possible actions, based on a set of chosen probabilities, such a strategy is called *mixed strategy*. The chosen probabilities in a mixed strategy for a player can be his/her own subjective probabilities that other players will choose certain strategies. It is a rational expectation if each player's belief about the other player's choices coincides with the choices that other players actually make. Therefore, Nash equilibrium is a certain kind of rational expectation equilibrium.

3.5.1 Nash Equilibrium

The concept of Nash equilibrium in a general or oligopolistic market is a bit convoluted at first. In oligopolistic markets, firms take strategic actions after considering the likely responses of other firms. Other firms will do the same, given the likely responses of other firms to their actions. If a firm is doing its best it can, given that other competing firms are also doing their best, market will reach an equilibrium point, which we call *Nash equilibrium*. Formally, Nash equilibrium represents a set of strategies or actions that each firm takes, given the best responses of its competitors' actions. Nash equilibrium gives us a basis for determining an equilibrium in an oligopolistic market. It is also the most commonly used solution concept in game theory.

Let N be a finite set of players, A_i be a set of actions available to player i ($i \in N$), notation $\succsim_i$ be a preference relation of player i on A. Formally, Nash equilibrium of a strategic game $\langle N, (A_i), (\succsim_i) \rangle$ is a profile $a^* \in A$ of actions with the property that for every player $i \in N$, we have

$$(a^*_{-i}, a^*_i) \succsim_i (a^*_{-i}, a_i) \quad \text{for all} \quad a_i \in A_i \tag{3.37}$$

The necessary condition for a^* to be a Nash equilibrium requires that no player i has an action that will yield an outcome that he prefers to the outcome that is generated when the player i chooses an action a_i^*, given that every other player j chooses his equilibrium action a_j^*. In other words, at Nash equilibrium, no player can profitably deviate away from his optimal action given the optimal actions of other players.

In general, the key assumption in Nash equilibrium is that each player knows his/her own strategies and payoffs as well as those of other players in a game. This assumption is known as *common knowledge*. It is also assumed that all players are rational, which means that each player will choose an action that will maximize his/her own payoff, given his/her subjective beliefs and that those beliefs are modified when new information arrives according to Bayes' law.

Nash (1950) has shown that an equilibrium will always exist for a game with a finite number of agents and a finite number of pure strategies. The equilibrium can also include mixed strategies. In general, games need not have a single Nash equilibrium. Games can have multiple Nash equilibria. Sometimes, there is no Nash equilibrium.

Games that are studied in game theory come in different forms and flavors. In contrast to one-shot game, *repeated games* are games in which actions are taken and payoffs are received over and over again. It can be *infinitely repeated games* or *games with finite repetitions*. For example, electricity markets are infinitely repeated games. In a repeated game, the strategy space for each player is much larger, as each player can choose a strategy as a function of the entire history of the game up until that point. A player, when making his/her choice, must take into account of the fact that other players can modify their behaviors based on his/her history of choices.

In reference to Prisoner's Dilemma game described earlier, it is in the long-term interest of both prisoners to try to reach to a solution of (cooperate, cooperate) (*cooperate* here means 'do not confess'). The previous Nash equilibrium (confess, confess) seems like a solution to a one-shot game. In a repeated Prisoner's Dilemma game with a finite number of repetitions, the Nash equilibrium is still (confess, confess) in every round. In an infinitely repeated game, Nash equilibrium is gravitated towards (cooperate, cooperate).

In general, games are described in two different forms: *strategic form* and *extensive form*. Strategic form provides a reduced summary of a game while the extensive form shows an extended description of a game.

The strategic form of a game can be defined by exhibiting a set of players $N = (1, 2, ..., n)$. Each player i ($i \in N$) has a set of strategies A_i from which the player i can choose an action $a_i \in A_i$, and a payoff function $\phi_i(a)$ that indicates the utility/profit

Follower

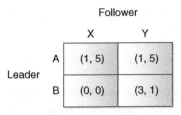

Figure 3.14 An Example Game.

that each player receives if a particular combination of strategies is chosen (where $s = (s_1, s_2, ..., s_n) \in S = \prod_{i=1}^{n} S_i$).

Sequential games are games in which players move in turn, responding to each other's actions and reactions. Stackelberg model is an example of a sequential game. In this model, the leader moves first before the followers take actions after the leader's move. Consider a game shown in Figure 3.14. It is easy to verify that the two Nash equilibrium in pure strategies are (A, X) and (B, Y). This is based on the assumption that both players make their choices simultaneously. Suppose that the leader moves first before the follower makes his/her choice. This game becomes a sequential game. Game tree is generally used to describe a sequential game. Game tree is a diagram that indicates the choices that each player can make at each point in time. The payoffs to each player are shown at the end of the tree leaves. Game tree is part of a description of the game in extensive form. This game tree is shown in Figure 3.15.

Once a choice has been made, for example after the leader's move, the follower is in a subgame consisting of strategies and payoffs available to them.

3.5.2 Bayesian Nash Equilibrium

For all of the games mentioned earlier, it is assumed that each player has complete knowledge about his/her strategies and payoffs as well as those of other players.

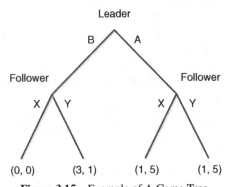

Figure 3.15 Example of A Game Tree.

Other players also know all such information. These are the games with complete information. However, this is not always the case in reality. Games with incomplete information can be approached slightly differently than games with complete information. In this type of game, all of the uncertainties about another player can be subsumed into player's type. The uncertainties of a player include the player's valuation of some good, risk profile, and strategies. Each player is assigned a player type and each player has some assumed probability distribution assigned over different types of players. A *Bayesian Nash equilibrium* of this game is a set of strategies for each type of player that maximizes the expected value of each type of player given the strategies taken by other players. This is similar to Nash equilibrium concept. However, Bayesian Nash equilibrium takes into account of additional uncertainty related to the type of players.

Suppose that there are N bidders in an auction, with valuations $(v_1, ..., v_N)$ and b_i represents the bid of bidder i. The beliefs of bidder i about other bidders' valuations are independent of v_i and they are the same for all bidders. Let us denote by V_{-i} the set of all possible combinations of valuations of all bidders except for bidder i. A Bayesian Nash equilibrium is a set of strategies $(s_1^*, ..., s_N^*)$ that specify the actions of the players given their own valuations and such that for each bidder i, the strategy $s_i^*(v_i)$ is on average the best response, assuming i's rivals play according to $(s_1^*, ..., s_N^*)$. In other words, $s_i^*(v_i)$ solves the maximization problem as

$$\max \sum_{v_{-i} \in V_{-i}} (\phi_i(b_i)|s^*(v_{-i}))Pr(V_{-i} = v_{-i}) \tag{3.38}$$

3.6 OLIGOPOLY

Oligopoly is a market in which only a few firms compete with one another, and entry by new firms is limited. Such a market is also known as oligopolistic market. Products may or may not be differentiated. Barriers to entry makes it difficult for new firms to enter the market, thus some or all firms in oligopolistic markets earn substantial profits. Oligopolistic market is also the most prevalent form of market structure.

Reasons for high barrier to entry include scale economies, patent, access to proprietary technology, and high investment cost. These barriers to entry are known as natural barriers to entry. Other reasons for barrier to entry can be the strategic actions taken by incumbent firms to deter entry by new firms. Such actions include dumping, predatory pricing, and building excess production capacity.

The monopoly power and profitability in oligopolistic markets depend in part on how firms interact. For example, if firms' interaction is more competitive than cooperative, prices would be closer to marginal cost, hence firms will earn lower profits. On the other hand, if firms are more cooperative, prices would be much higher than marginal cost, and thus firms can earn higher profits.

In some oligopolistic markets, firms do compete aggressively, but in others, they do cooperate. The key question is how oligopolistic firms decide on output and prices.

The key point in answering this question lies in the strategic interaction and decisions made by firms. It means that each firm in an oligopolistic market has to consider the probable actions and reactions of its competitor firms. Therefore, understanding the basic concepts of gaming (game theory) and strategy is key to understand oligopolistic markets.

In oligopolistic markets, because only a few firms are competing, each firm must carefully consider how its actions will affect its rivals, and how its rivals are likely to react. One possible strategic action for a firm would be to reduce the price of its products. As a consequence, price war, in which the price of a product falls drastically, is one possible outcome of strategic actions and responses in an oligopolistic market. So, a firm needs to weigh all possible outcomes of its strategic action. Changing production level is another possible strategic action that a firm can take. These strategic considerations are complex because other firms will also be in the same position in the sense that they must consider carefully the responses of other firms to their strategic decisions and actions.

There are four basic models used to study oligopolistic markets: *Cournot model*, *Stackelberg model*, *Bertrand model*, and *Collusion*.

3.6.1 Cournot Model

Cournot model, introduced by a French economist Augustin Cournot in 1838, is an oligopoly model in which all firms decide simultaneously how much to produce while treating the output of its competitors fixed. Assume that the market under consideration is duopoly where there are only two firms competing each other. Also assume that firms in this duopolistic market produce homogeneous good and know the market demand curve. Then each firm must decide how much to produce, and the two firms make their decisions at the same time. When making its production decision, each firm takes its competitor's output into account. The competitor firm is also deciding how much to produce at the same time. So market price will depend on the total output of both firms.

A simple numerical example below will help illustrate the concept behind Cournot model. Assume that the two firms, firm 1 and firm 2, are competing each other in this duopoly. Assume the following costs of production for each firm:

$$c_1 = 20q_1$$
$$c_2 = 20q_2 \qquad (3.39)$$

In other words, each firm has constant marginal cost of $20 per unit. Let market demand function be given by $P = 200 - q_1 - q_2$. With this demand curve, the market price will fall if either firm tries to increase the amount of output it sells. The key question is how much each firm will produce given their cost functions and demand function. Therefore, in Cournot model, each firm will try to make a guess about how much its rival will produce and believe that its rival will stick to that estimated output level. Each firm's optimal level of output is the best response to the output level that it expects that its rival will choose. In other words, each firm tries to choose an output level that maximizes its own profit, given the output it thinks its rival will choose. Hence, the optimal output $q_1{}^*$ for firm 1 will depend on the optimal output $q_2{}^*$ of firm 2 that it thinks its rival will choose.

Thus for firm 1, the profit to be maximized is

$$\pi_1 = R_1(q_1) - c_1 = P(q_1)q_1 - c_1 = (200 - q_1 - q_2)q_1 - 20q_1 \qquad (3.40)$$

The optimal output for firm 1 that maximizes its profit can be found by using profit-maximization principle:

$$q_1{}^* = 90 - 0.5q_2 \qquad (3.41)$$

This profit-maximizing value of $q_1{}^*$ is the best response of firm 1 given different output levels of firm 2. Equation (3.41) is also called firm 1's *reaction function*. Similarly, the best response of firm 2 to the different output levels of firm 1 can be found as

$$q_2{}^* = 90 - 0.5q_1 \qquad (3.42)$$

In equilibrium, each firm's guess about its rival's output must be correct. Each firm will adjust its output until each firm's guess about other firm's output is correct, hence there is no more incentive to change its output. We can find this optimal pair of output $(q_1{}^*, q_2{}^*)$ for both firms by solving both firms' reaction functions simultaneously. The solution turns out to be $q_1{}^* = q_2{}^* = 60$, which are the optimal output for both firms. In Figure 3.16, this equilibrium output corresponds to the point of intersection of two reaction functions. The equilibrium price is found to be $80 and each firm will make profit of $3600.

Cournot equilibrium, sometimes called *Cournot–Nash equilibrium*, is an equilibrium in Cournot model, in which each firm *correctly* assumes how much its competitor will produce and sets its own production level that maximizes its own profits. As a consequence, no firms have incentive to deviate away from this equilibrium.

Cournot model and Cournot equilibrium will not hold if any firm starts to change its output. This change in output runs counter to the assumption of fixed output by the competing firms in the Cournot model. Therefore, Cournot model applies for a non-repeating or one-shot game.

3.6.2 Stackelberg Model

Alternative model of oligopolistic behavior can be represented by *Stackelberg model* introduced by German economist Heinrich Freiherr von Stackelberg in 1934. Stackelberg model is an oligopoly model in which one firm sets its output first before other firms do. The first firm which decides its output acts as a *leader* while other firms act as *followers*.

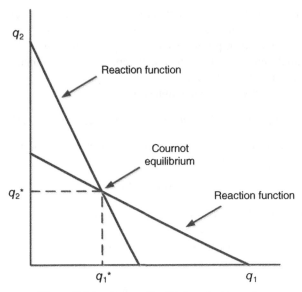

Figure 3.16 Cournot Equilibrium for Duopoly.

Consider a case of duopoly. In this market, assume that firm 1 is a leader while firm 2 is the follower. The leader may be a larger firm or a firm with significant competitive advantage. Firm 2 is waiting for firm 1 to make the first move. Firm 2 will choose its output level q_2 given the output q_1 of firm 1, using the reaction function mentioned in Cournot model. Thus, firm 2 will choose output q_2 that will maximize its profit:

$$\pi_2 = p(q_1 + q_2)q_2 - c_2(q_2), \tag{3.43}$$

where $p(q_1 + q_2)$ is the inverse market demand function. From this, we can derive the firm 2's reaction function $Q_2(q_1)$ which provides the best response of firm 2 to the different output levels of firm 1. By the same token, firm 1 also accounts for the reaction of firm 2 when choosing its output q_1. Since firm 1 is the leader, it will therefore choose output q_1 to maximize its profit:

$$\pi_1 = p(q_1 + Q_2(q_1))q_1 - c_1(q_1) \tag{3.44}$$

Once the profit function for firm 1 is derived using only one variable q_1, the optimal output $q_1{}^*$ for firm 1 that will maximize its profit can be solved directly by using calculus. Using the same duopoly example given in the Section 3.6.1, the optimal output for firm 1 turns out to be $q_1{}^* = 90$ and that for firm 2 is $q_2{}^* = 45$. The equilibrium price is found to be \$65 and firm 1's profit is \$4050 which is more than the profit that firm 1 makes in Cournot duopoly model, while firm 2's profit is just \$2025 which is much less than the profit that firm 2 makes in Cournot duopoly model. The actual calculation for these results is left as an exercise to the students.

In general, the leader firm does better in Stackelberg equilibrium than in Cournot equilibrium.

3.6.3 Bertrand Model

We have seen that in both Cournot and Stackelberg models, firms use quantity output as their strategic variable. In contrast, if price is chosen as the strategic variable, the outcome of the market can be quite different. In *Bertrand model*, introduced by another French economist Joseph Bertrand in 1883, firms compete using price rather than quantity output as their strategic variables in markets with homogeneous products. Bertrand model is an oligopoly model in which firms produce a homogeneous product, each firm treats the price of its competitors as fixed and all firms decide simultaneously what price to charge, but firms cannot cooperate. Like in Cournot model, firms make their decisions at the same time. With just two firms in a market (duopoly), Bertrand model produces outcome as in a perfectly competitive market as illustrated below.

Let us return to our duopoly example given in the previous section on Cournot model, in which the market demand curve is given by $P = 200 - q_1 - q_2$, where the sum of q_1 and q_2 will be the total production of a homogeneous good. Again, each firm has constant marginal cost of $20 per unit. Now suppose that the two duopolists compete by simultaneously choosing a price instead of a quantity. Because the good is homogeneous, consumers will purchase only from the lowest-price seller. Therefore, if one firm charges a price that is higher than the price charged by the other firm, the lower-price firm will supply the entire market while the higher-price firm will sell nothing. If both firms charge the same price, consumers will be indifferent as to which firm they will choose to buy from, so each firm will equally supply half of the market.

The best bet for each firm is to charge a price equal to its marginal cost because each firm knows neither the price that the other firm would charge nor the marginal cost of that firm. In this particular case, since the marginal costs of both firms are equal, the price set by both firms will be $20. As a consequence, the optimal output for each firm will be $q_1{}^* = q_2{}^* = 90$. Surprisingly, because each firm charges a price which equals to its marginal cost, the profit that each firm will earn is zero. Again, this is consistent with the outcome of a market in a perfect competition, that is firms make zero profit in a perfectly competitive market. This is also equivalent to *Nash equilibrium* outcome. While there are some criticisms in terms of realistic outcome in Bertrand model, it is a useful model because it shows how the equilibrium outcome in an oligopoly depends on the firms' choice of strategic variables.

3.6.4 Collusion Model

All of the models previously described are examples of *noncooperative games*. In other words, firms in previous models do compete by making decisions that will maximize their profits, independently of each other. In noncooperative games, negotiations and enforcement of binding contracts among firms are not possible. The natural question is what would happen to the market outcome if the firms cooperate? A market structure in which firms *explicitly agree to cooperate or collude* in setting prices and output levels is called a *cartel*.

Cartel is a market in which some or all firms collude explicitly. Cartels can be formed if only a subset of production firms agree to cooperate. But if sufficient number of firms adhere to cartel's binding agreement and if the market demand is sufficiently inelastic, the prices set by the cartel can be well above competitive levels. However, firms in a cartel also need to worry about the reactions of firms which are not part of the cartel. Also, it can be problematic for cartel if one or more of its members tend to violate their binding agreements.

Using the simple duopoly model again, consider the profit maximization problem of two firms in a cartel as

$$\max \quad \pi(q_1, q_2) = p(q_1 + q_2)(q_1 + q_2) - c_1(q_1) - c_2(q_2) \tag{3.45}$$

The first-order optimality conditions for this objective function will be

$$p'(q_1^* + q_2^*)(q_1^* + q_2^*) + p(q_1^* + q_2^*) = c_1'(q_1^*)$$
$$p'(q_1^* + q_2^*)(q_1^* + q_2^*) + p(q_1^* + q_2^*) = c_2'(q_2^*) \tag{3.46}$$

Thus, the profit maximization implies that $c_1'(q_1^*) = c_2'(q_1^*)$. One can see that the cartel solution is not stable unless the two firms are merged. For each firm, there is an incentive to undercut the other firm by producing more than the agreed-upon output (quota). If firm 2 will hold its output constant, the first-order condition for firm 1's profit maximization can be rearranged as

$$\frac{\partial \pi_1(q_1^*, q_2^*)}{\partial q_1} = p'(q_1^* + q_2^*)(q_1^*) + p(q_1^* + q_2^*) - c_1'(q_1^*)$$
$$= p'(q_1^* + q_2^*)(q_2^*) > 0 \tag{3.47}$$

If firm 1 assumes that firm 2 will produce the output that is agreed-upon, then firm 1 will have an incentive to produce more than its agreed-upon output. Even if the firm 2 produces more than its quota, it is still profitable for firm 1 to produce more than its quota. This incentive for firms in a cartel to deviate away from the cartel's agreement makes the cartel solution unstable.

Table 3.1 depicts a summary of the main aspects of the different types of markets.

Table 3.1   Market Type Comparison

Market Type	Monopoly	Oligopoly	Monopolistic Competition	Perfect Competition
Producers	One	Few	Many	A great many
Product	Unique	Similar	Differenciated	Identical
Barriers to entry	High	Relatively high	Easy	No barriers
Competition	None	Non-price	Non-price; price	Price
Price type	Price setter	Price maker; independent behaviour	Price maker	Price taker

3.7 AUCTION THEORY

Theory of auction explains the structures and the rules associated with auction markets and the possible market outcomes (revenues, efficiency, etc.). Auction markets are markets in which products are bought or sold through formal bidding processes. Advantages of auctions include (1) less time-consuming than one-on-one bargaining, (2) encouragement of competition among buyers in a way that increases the seller's revenue, and (3) low cost of transactions.

Auctions are inherently interactive, with many buyers competing to obtain an item of interest. This interaction in an auction can be particularly valuable for sales of items that are unique and do not have established market value. It can also be helpful for the sale of items that are not unique but whose value fluctuates over time. Choice of auction format can affect the seller's auction revenue. There are several different kinds of auction formats.

1. **Traditional English (or Oral) Auction**: An auction in which a seller starts by offering an item with a very low price. Then, the seller actively solicits progressively higher bids from a set of potential buyers. As bidding progresses, the bid price rises continuously and exogenously while bidders continue to show their willingness to buy at the prevailing price. As the auction progresses, some bidders start to exit when the current bid price is greater than their valuations. The auction ends when there is only one remaining bidder. This bidder wins the auction and pays a price equal to that at which the auction stops, which is the exit price of his last opponent. In other words, the auction ends with the second-highest bid. English auction is also known as ascending auction. This type of auction is commonly used for selling goods, such as antiques and artwork.

2. **Dutch Auction**: An auction in which a seller begins by offering an item at a relatively high price, then reduces it by fixed amounts. The auction ends when one bidder accepts the price. In this kind of auction, only one bid, that

is, the winning bid, can be observed. Dutch auction is the converse of English auction and also known as descending auction.

3. **Sealed-Bid Auction**: An auction in which all bids are submitted simultaneously in sealed envelopes. The winning bidder is the individual who has submitted the highest bid. There are two variants in this type of auction. The first variant is *first-price sealed-bid auction* in which bids are submitted simultaneously, and good is awarded to the highest bidder at a price equal to his bid. The second variant is *second-price sealed-bid auction* in which bids are submitted simultaneously, and good is awarded to the highest bidder at a price equal to the price of the second-highest bidder. This type of auction is also known as *Vickrey's auction*, created by economist William Vickrey. A modification of this type of auction is used by eBay's system and online advertisement programs at Google and Yahoo.

The types of auctions described earlier apply to auctions for single items. In the case of auctions for homogeneous goods, each of the M individual items is identical or close substitute so bids can be expressed in terms of quantities without the need to indicate the identity of the particular good that is being auctioned. Auction rules and bidding can be much simpler when homogeneous goods are auctioned. Recall that the electricity product in an electricity market is treated as homogeneous good. There are three kinds of sealed-bid, multiunit auction formats for M homogeneous goods.

In each of these, a bid comprises an inverse demand function, that is, a (weakly) decreasing function $p_i(q)$, for $q \in [0, M]$, representing the price offered by bidder i for a first, second, etc., unit of the good. The bidders submit bids; the auctioneer then aggregates the bids and determines a clearing price. Each bidder wins the quantity demanded at the clearing price, but his payment varies according to the particular auction format.

1. **Pay-As-Bid Auction**: Each bidder wins the quantity demanded at the clearing price, and pays the amount that he bid for each unit won. Pay-as-bid auctions are also known as *discriminatory auctions* or *multiple-price auctions*.

2. **Uniform-Price Auction**: Each bidder wins the quantity demanded at the clearing price, and pays the clearing price for each unit won. Uniform-price auctions are also known as *nondiscriminatory auctions*, *competitive auctions*, or *single-price auctions*. Most of the electricity markets in many countries are based on uniform-price auction.

3. **Multiunit Vickrey Auction**: Each bidder wins the quantity demanded at the clearing price, and pays the opportunity cost (relative to the bids submitted) for each unit won.

3.8 FURTHER DISCUSSIONS

In this chapter, we have provided some introduction to microeconomic theories which are essential for understanding the complex operation of electricity market and related issues. Most students who will want to study each of the key topics in this chapter in more detail can further explore from additional references. Application of which theory or model described in this chapter in the context of electricity market depends on the objective and nature of the problem of interest. For example, theory on social welfare (sum of consumer surplus and producer surplus) is useful when answering the question on the effect of any system change, such as adding a new transmission line, on the system-wide economic benefit. Some researchers will be interested in whether a particular part of the market can reach an equilibrium, if any. Antitrust officials or independent market monitors will be interested in whether a market player has market power and is exercising it. If it does, what are the effects on the social welfare? A research question can be posed on whether a uniform-price auction or pay-as-bid auction is better in improving or maximizing the economic efficiency of a particular electricity market. There are other more complex issues that the reader can relate to after having some level of understanding of the theories in this chapter.

CHAPTER END PROBLEMS

3.1   A supply curve is generally modeled as upward sloping curve. Is it possible, in reality, for a supply curve to have a downward sloping curve like a demand curve? Why and why not?

3.2   What would happen to the supply curve if the cost of a raw material to produce a product suddenly increases? What would happen to the demand curve if the income of a consumer decreases?

3.3   What can be a substitutable product for electricity?

3.4   What is the nature of price elasticity of demand for electricity? Do you think consumers will reduce their electricity consumption when electricity price increases suddenly? Will that elasticity now be the same in next 5 years? What triggers that change, if any?

3.5   Suppose the supply curve of a market product is given by $Q_s = 1800 + 240P$, while its demand curve is given by $Q_D = 3550 - 266P$. What are the market-clearing price and market-clearing quantity for this product? What are the price elasticities of supply and demand at this price and quantity?

3.6   Give some examples of normal good and inferior goods. Do you think electricity is a normal good?

3.7   Is electricity demand for industrial customers elastic or inelastic? Why? How about for commercial and residential consumers?

3.8   Suppose there are only two customers for a product in a market. The demand curve of each consumer is given as $Q_1 = 1465 - 88P$, and $Q_2 = 1344 - 138P$. Compute the final market demand curve. Also draw demand curves for both customers and market.

3.9   Explain the difference between fixed cost and sunk cost. Give an example of each and explain why they are different.

3.10   Is electricity a homogeneous product? Can we differentiate electricity based on product quality?

3.11   Suppose the production costs for two firms are given by $c_1 = 10q_1$, $c_2 = 8q_2$. Also suppose market demand be $P = 100 - q_1 - q_2$. Find the optimal output and profit for each firm, as well as equilibrium price (a) using Cournot model and (b) using Stackelberg model.

3.12   Give an example of cooperative game and noncooperative game. Why they are possible?

3.13   Consider the following coordination game (Figure 3.17):

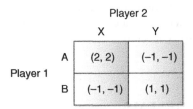

Figure 3.17   An Example Game for Problem 3.13.

   a. Calculate all the pure strategy equilibria of this game.
   b. Do any of the pure strategy equilibria dominate any of the others?
   c. Suppose *Player 1* moves first and commits to either strategy A or strategy B. Are the strategies you chose above still Nash equilibria?

3.14   In the energy market part of an electricity market, generators (suppliers) have to bid in to be able to serve system demand. This market clears using an auction mechanism. Generators with the lowest bids are generally selected for this opportunity to serve demand. Assuming there is no congestion, all cleared generators are paid, and all demand consumers pay the same market-clearing price. The market clearing in energy market is what type of auction mechanism?

FURTHER READING

1. Varian HR. *Intermediate Microeconomics: A Modern Approach*, 8th edition. W. W. Norton & Company; 2010.

2. Varian HR. *Microeconomic Analysis*. W. W. Norton & Company; 1992.

3. Frank RH. *Microeconomics and Behavior*. McGraw-Hill; 1994.

4. Pindyck RS, Rubinfeld DL. *Microeconomics*. Pearson Prentice Hall; 2005.

5. Besanko D, Dranove D, Shanley M, Schaefer S. *Economics of Strategy*. John Wiley & Sons; 2007.

6. Neumann JV, Morgenstern O. *Theory of Games and Economic Behavior*. Princeton University Press; 2004.

7. Tirole J. *The Theory of Industrial Organization*. The MIT Press; 1988.

8. Fudenberg D, Tirole J. *Game Theory*. The MIT Press; 1991.

9. Shepherd WG, Shepherd JM. *The Economics of Industrial Organization*. Waveland Press; 2004.

10. Osborne MJ, Rubinstein A. *A Course in Game Theory*. The MIT Press; 1994.

11. Bierman HS, Fernandez L. *Game Theory with Economic Applications*. Addison-Wesley; 1998.

12. Kreps DM. *Game Theory and Economic Modelling*. Oxford University Press; 2001.

13. Krishna V. *Auction Theory*, 2nd edition. Academic Press; 2009.

14. Ruiz C, Conejo AJ, Smeers Y. Equilibria in an oligopolistic electricity pool with stepwise offer curves. *IEEE Transactions on Power Systems* 2012;27(2):752–761.

Chapter 4

Power System Unit Commitment

Starting from this chapter towards next two chapters, we will introduce the key components of economic operation of a power system. These key components are unit commitment (UC), economic dispatch, and optimal power flow. These topics are intricately related. The solutions from the unit commitment and economic dispatch based on a particular OPF determine the optimal generation schedule. In an electricity market setting, generator bids, in place of their costs, are generally used to determine that optimal outcome. Readers are suggested to follow each of these three chapters to have a fundamental understanding of power system economic operation.

4.1 INTRODUCTION

On a high level, power system economic operation concerns the operation of generation, transmission, and distribution systems in the least-cost manner with the goal of achieving instantaneous balance of generation and load. At the same time, the voltage at every node in the system and system frequency should be within acceptable limits. Achieving the twin goals of improving not only the economic aspect of the electric system but also the technical security in the most efficient way is one of the most important challenges for the system operator. Both economic and operational efficiencies are so important that they are sometimes referred to as the hidden fuel. That is one of the reasons why there has been an increase in the investment in energy efficiency developments which nowadays is comparable to the investments in other energy related areas such as renewable electricity. This development can be observed in many countries and all electricity domains.

Electricity Markets: Theories and Applications, First Edition. Jeremy Lin and Fernando H. Magnago.
© 2017 by The Institute of Electrical and Electronics Engineers, Inc. Published 2017 by John Wiley & Sons, Inc.

To achieve this near balance of instantaneous generation and load in the least possible cost, system operators have to undertake important tasks such as unit commitment, economic dispatch, and optimal power flow. These topics will be covered in this and next two chapters. Unit commitment is a large-scale, non-convex, mixed-integer optimization problem which requires effective and efficient numerical methods to obtain its solution. Economic dispatch is a step that needs to be done after completing the unit commitment process while the *optimal power flow* (OPF) solves the economic dispatch by adding additional constraints to the optimization problem. In all these activities, the optimization methods from the operations research field play a significant role in the decision-making process associated with both unit commitment and economic dispatch.

Formulating an optimization problem, particularly for a large-scale system, is not easy because algorithms and models are complicated. There is no universal optimization algorithm that can solve all optimization problems. And it is the responsibility of the user to choose an algorithm that is appropriate for a particular application. The selected algorithm largely determines the performance and accuracy of the solution. Once the algorithm is implemented and applied to a specific model, it is important to know whether it successfully finds an optimal solution or not. For example, if there is no optimal solution, it is also important to know whether additional information is available in order to improve the suboptimal solution. Moreover, a sensitivity analysis is generally needed to develop the model. At minimum, an optimization problem formulation must set the objective function which measures the idea of optimality, explicitly state the constraints that must be enforced, and make the solution feasible. A particular problem can be described in many different forms. In general, the mathematical formulation of an optimization problem can be represented by Eq. (4.1):

$$\text{minimize} \quad f(\mathbf{x}) \tag{4.1}$$

$$\text{subject to}$$
$$h_i(\mathbf{x}) = \mathbf{0}, \quad i = 1, ..., r$$
$$g_j(\mathbf{x}) \le \mathbf{0}, \quad j = 1, ..., p$$
$$\mathbf{x} \quad \epsilon \quad \Omega,$$

where h_i and g_j are given functions; $f, h_i, g_j \in C^1(\Omega)$ are continuous functions. The objective is to find $\mathbf{x}^*$ such as

1. $\mathbf{x}^* \epsilon \Gamma = \Omega \cap \{\mathbf{x} \epsilon R^n : h_i(\mathbf{x}) = \mathbf{0}, g_j(\mathbf{x}) \le \mathbf{0}\}$, where Γ is the feasible region.
2. $\forall x \epsilon \Gamma : f(\mathbf{x}^*) \le f(\mathbf{x})$

A particular type of optimization problem is the convex optimization in which both the objective and constraint functions are convex:

$$f(\alpha x + \beta y) \leq \alpha f(x) + \beta f(y) \tag{4.2}$$

if $\alpha + \beta = 1$, $\alpha \geq 0$, $\beta \geq 0$

In the context of market clearing in an electric market, depending on the problem that needs to be solved, the models in question can have different formulations. In this chapter, our focus is on the specific optimization methods that are directly relevant and applicable to solving optimization problems in areas related to the operation of both power system and electricity market. Our goal is to expose the readers to these optimization methods along with their problem formulations and solution methods. Different optimization methods are used in different aspects of the power system and market operation. The fundamental formulations with respect to the unit commitment problem are described next.

4.2 UNIT COMMITMENT

As stated earlier, the load of a power system experiences a cyclical nature that depends on the time of a day, the type of day, and the season. The installed generation capacity must provide for the safe supply of the peak demand. Therefore, there is an idle capacity of available generation at non-peak periods. During the operational planning, typically a week before the actual system operation, decisions need to be made on which generation unit must be in service (availability) and the amount of active power they need to provide (scheduling). The economic consequence of this task is crucial because it is very expensive to turn on a generating unit which is in addition to the operating costs of the different generating unit technologies. Furthermore, there are different constraints related to the system and generating units that condition the numbers and types of generating units that can be used to serve the demand. As a consequence, the market and system operator faces a complex problem of scheduling the generators for the economic operation of the system. The time horizon for the generation scheduling can extend from 24 hours of a day as in near real-time operation up to several years during planning studies. The formal definition of the unit commitment problem can be stated as

Problem Definition: *Considering a set of generation units available and a prede-fined demand profile for a particular scheduling period, find for each time interval the subset of generating capacity that meets the required demand at a minimum operating cost and subject to operating constraints.*

Mathematically, the unit commitment problem is a dynamic optimization problem since it establishes the optimal operation policy over a scheduling horizon. The solution procedure involves an economic dispatch as a subproblem because the generation units must be committed in the most economical way at each time interval. The objective function can be stated as the minimization of the system operational cost which may include

1. Generation units cost
2. Start-up and shut down cost
3. Maintenance cost

The constraints can be grouped into three categories: system, plant, and unit constraints. Examples of these type of constraints are given below.

1. **System**: power balance, reserves, fuel constraints
2. **Plant**: crew, emissions, fuel constraints
3. **Unit**: power limits, minimum time on or off, transition ramp constraints

4.3 MATHEMATICAL FORMULATION FOR UNIT COMMITMENT

Mathematically, the UC problem can be represented as a mixed integer programming (MIP) problem. Moreover, if the problem can be linearized, the problem can be represented as a mixed integer linear programming (MILP) problem. An MILP is a linear programming (LP) problem in which some variables are integers. If all variables are binary variables, then the problem is known as 0-1 MILP. If all variables are integers, the problem is known as linear integer programming problem (IP).

Binary variables allow the modeling of certain nonlinearities which are common in electricity market problems and difficult to include. Some examples of such nonlinearities are conditional constraints, discontinuous functions, or piecewise non-convex functions.

Mathematically, an MILP problem can be formulated as an LP problem. However, considering that some of the variables are integers, then by repeating the LP formulation, the MILP problem is

$$\text{minimize} \quad f(x) = c^T x \tag{4.3}$$

subject to

$$Ax = b$$
$$x \geq 0$$
$$x_j \quad \text{are integers for some} \quad j = 1, 2, ..., n$$

Considering this general formulation, the UC problem can be represented mathematically. From the system operator perspective, the UC objective function is to minimize the sum of the production costs of all generation units and the associated startup costs which can be represented by Eq. (4.4). On the other hand, from the generator owners' perspective, the UC objective function is the addition of the payments of power, energy, and ancillary services and can be represented by Eq. (4.5).

$$\min \sum_{t=1}^{T} \sum_{g=1}^{G} \left(Cp_{gt} + Ca_{gt} \right) \tag{4.4}$$

$$\max \sum_{t=1}^{T} \sum_{g=1}^{G} Pr_{gt}, \tag{4.5}$$

where T is the total simulation time, G is the total number of units, Cp_{gt} is the production cost of unit g at period t, Ca_{gt} is the start-up cost of unit g at period t, and Pr_{gt} is the payment to unit g at period t.

In terms of modeling the constraints, the power balance equation is

$$\sum_{g \in G} p_{gt} = D_t \qquad \forall t, \tag{4.6}$$

where D_t is the demand at period t and p_{gt} is the active power generation of unit g at period t. The spinning reserve can be modeled as

$$\sum_{g \in G} rr_{gt} \geq Rr_t \qquad \forall t, \tag{4.7}$$

$$\text{for } rr_{gt} = \min \left\{ u_{gt} \left(TMSR_g \cdot MSR_g \right), \left(\overline{P}_g - p_{gt} \right) \right\}$$

where $\overline{P}_g$ is the maximum active power capacity of unit g at period t; Rr_t is the required reserve margin at period t; rr_{gt} is the variable related to the spinning reserve of unit g at period t; MSR_g is the maximum $\Delta P/\Delta T$ that unit g can afford; $TMSR_g$ is the time ΔT that unit g can provide ΔP, and u_{gt} is the status of unit g at period t.

The operational reserve is defined as follows:

$$\sum_{g \in G} ro_{gt} \geq Ro_t \qquad \forall t, \tag{4.8}$$

$$\text{for } \quad ro_{gt} = \begin{cases} u_{gt} \, qsc_g & \text{unit off} \\ rr_{gt} & \text{unit on} \end{cases}$$

where ro_{gt} is the operational reserve of unit g at period t; Ro_t is the operational reserve margin required by the system at period t; qsc_g is the start-up capacity of unit g.

Another system constraint is the fuel constraint which can be represented as

$$\underline{\text{Fuel}} \leq \sum_{g \in G_{\text{fuel}}} \sum_{t \in T} u_{gt} \, \text{Cf}_{gt} \leq \overline{\text{Fuel}}, \tag{4.9}$$

where $\underline{\text{Fuel}}$ and $\overline{\text{Fuel}}$ are the minimum and maximum system fuel limits; G_{fuel} is the subset of units that share the same fuel type; Cf_{gt} is the fuel consumption of unit g at period t.

In addition to these constraints, temporal constraints are also considered for a UC problem. For example, the minimum service time and minimum out of service time can be formulated as

$$(x_{g,t-i} - Mut_g)(u_{g,t-1} - ug, t) \geq 0 \tag{4.10}$$

$$(x_{g,t-i} - Mdt_g)(u_{g,t} - ug, t - 1) \leq 0 \tag{4.11}$$

and the technical limits of the generator units become

$$u_g^t P_g^{\text{Min}} \leq P_g^t \leq u_g^t P_g^{\text{Max}} \tag{4.12}$$

for all periods $t = 1, ..., T, i = 1, ..., M$; where u_g^t is the status of unit g at period t, Mut_g, and Mdt_g are the minimum up time and minimum down time, respectively for unit g, and P_g^{Min}, P_g^{Max} the active power limits of unit g.

The mathematical formulation of these equations may change depending on the solution method selected. In addition, the constraints included here are for illustrating a general UC formulation. In a practical application, the number of constraints may change depending on the specific problem associated with the electric system and the operator needs.

4.4 NUMERICAL METHODS FOR UNIT COMMITMENT PROBLEM

Lagrange relaxation (LR) method is one of the prevailing methods applied to solve UC problems. One of the main advantages of the LR method is the possibility of decoupling the generating units with respect to the power balance equations and all other system constraints. In addition, it allows to calculate the dual multipliers (marginal cost) directly. LR method has different versions: subgradient, interior point, and augmented LR. Also, many research works incorporate different constraints into LR formulation such as ramp constraints, fuel constraints, crew constraints, network constraints, and security constraints. Primal based methods are also proposed: branch and bound, linear programming, dynamic programming, exhaustive enumeration, and priority list methods. Nowadays, the other popular technique is MIP, thanks to the advances in software tools and computing capacity. Many examples of the implementation of MIP for UC problems can be found in the literature. These two methodologies—LR and MIP—are widely used to solve UC problems. LR methodology is still applied for this problem particularly when the simulations require larger number of periods (e.g., long term studies).

Finally, heuristic methods are also proposed for solving UC problems. The solution methods in this group are based on genetic algorithm, simulated annealing, evolutionary programming, fuzzy logic, artificial neural networks, and Tabu search. Additionally, there are alternative proposals based on ant colony and network flow methods.

4.4.1 Heuristic Methods

4.4.1.1 Exhaustive Enumeration
The UC problem can be solved by counting all possible combination of generation units. Considering a set of M generating units and a simulation time frame T, then the total number of combination is $(2^M - 1)^T$. Most of the combinations are infeasible due to the constraints, and thus the number of feasible combinations is drastically reduced. The disadvantage of this method is related to the dimension of the solution space. Therefore, it is only applicable to small systems.

Example: Let us consider a small system with three generators available and a 500 MW load. The hourly consumption can be represented by a second order polynomial. The polynomial coefficients and the units limits are described in Table 4.1.

The total number of combinations is 2^3 and is described in Table 4.2.
The combination is feasible if the following constraint is not violated.

$$\sum_i P_{G_i}^{\text{Min}} \leq P_D \leq \sum_i P_{G_i}^{\text{Max}} \tag{4.13}$$

Please note that for simplicity, in this example, losses are not considered. That is the reason that the combination $(G_1 + G_3)$ is feasible, otherwise is unfeasible. The dispatch for the three feasible solutions are the ones described in Table 4.3. Last combination is the one that must be selected since it gives the minimum consumption.

Table 4.1 Hourly Consumption Data

ID	Type	a [kcal/h]	b [kcal/MWh]	c [kcal/MW^{2}h]	P^{Min} [MW]	P^{Max} [MW]
G_1	TV	17,787	2227	9.60	150	300
G_2	TV	13,387	2373	14.20	150	250
G_3	TV	14,931	2500	18.95	100	200

Table 4.2 Possible Combinations

G_1	G_2	G_3	Condition
Off	Off	Off	Unfeasible
Off	Off	On	Unfeasible
Off	On	Off	Unfeasible
Off	On	On	Unfeasible
On	Off	Off	Unfeasible
On	Off	On	Feasible
On	On	Off	Feasible
On	On	On	Feasible

4.4.1.2 Priority List This is the simplest heuristic method. An ordered list based on the marginal fuel consumption at maximum capacity is conformed. The heuristic methods, in general, can be computationally efficient. However a global optimal solution is difficult to achieve.

Example: For the same system considered in the previous example, the marginal consumptions are

$$\frac{dC_{G_1}}{dP_{G_1}} = 2227 + 2 \times 9.60 P_{G_1}^{\text{Max}} = 7987 (\text{kcal/MWh})$$

$$\frac{dC_{G_2}}{dP_{G_2}} = 2373 + 2 \times 14.20 P_{G_2}^{\text{Max}} = 9473 (\text{kcal/MWh})$$

$$\frac{dC_{G_3}}{dP_{G_3}} = 2500 + 2 \times 18.95 P_{G_3}^{\text{Max}} = 10,080 (\text{kcal/MWh})$$

Based on the marginal consumption, the priority list is built, this list is illustrated in Table 4.4.

Ignoring constraints such as minimum time on or off and start-up, and considering the same load as before (500 MW), then this load is supplied by $G_1 + G_2$ based on the merit list. It is observed that this is not the most economic choice.

Table 4.3 Results of Exhaustive Enumeration UC

G_1(MW)	G_2(MW)	G_3(MW)	Total Consumption (kcal/h)
300	—	200	2,822,818.0
300	200	—	2,605,874.0
234.7	15.6	111.7	2,312,789.06

Table 4.4   Priority List

Combination	Minimum MW	Maximum MW
$G_1 + G_2 + G_3$	400	750
$G_1 + G_2$	300	550
G_1	150	300

4.4.2 Dynamic Programming

The dynamic programming technique was initially developed to solve numerical problems related to automatic control and dynamic optimization. Dynamic programming searches the optimal trajectory within the solution space. This search can be done backward or forward time frame. Each time period is named *stage* and the unit combination at each stage is known as *state*.

The minimum cost C_{min} at period K with the combination I is

$$C_{min}(K, I) = \min_{\{L\}}[C_{P_G}(K, I) + C_{tr}((K - 1, L) \Rightarrow (K, I)) + C_{min}(K - 1, L)],$$

(4.14)

where $C_{min}(K, I)$ is the minimum cost at state (K,I), $C_{P_G}(K, I)$ is the generation cost at state (K,I), and $C_{tr}((K - 1, L) \Rightarrow (K, I))$ is the transition cost from state $(K -1, L)$ to state (K,I).

The state (K, I) is the combination I of generation units at period K. The transition from one combination to another combination is known as *strategy*. Given the following definitions:

X is the number of combination to reach at each stage, and **N** is the number of strategies at each stage. The algorithm can be summarized as follows:

Start

- $K = 1$
- Calculate for each **X** combination at the first stage:

$$C_{min}(1, I) = C_{P_G}(1, I) + C_{tr}((0, L) \Rightarrow (K, I))$$

Cycle

- $K = K + 1$
- $\{L\}=$ set of N feasible combination at period $K- 1$
- Calculate the minimum cost of state (K,I) for the X combinations at stage K:

$$C_{min}(K, I) = \min_{\{L\}}[C_{P_G}(K, I) + C_{tr}((K - 1, L) \Rightarrow (K, I)) + C_{min}(K - 1, L)]$$

- Save the N strategies with lower cost.

- if K is the last period, store the optimal solution and finalize the algorithm, otherwise restart the cycle.

4.4.3 Dual Methods: Lagrange Relaxation

Dual methods consider the Lagrange multipliers as the main variables. Once they are determined, the optimal solution is calculated. The multipliers indicate the sensitivity of the objective function with respect to small variations in the constraints. Therefore, they are interpreted as the marginal costs associated with the constraints. These methods do not solve the optimization problem directly. However, they solve an alternative problem known as *dual problem*. To illustrate the generalities of these methods, consider the constrained optimization problem previously defined by Eq. (4.1). This problem is known as *original or primitive problem*. If $\mathbf{x}^*$ represents a local optimal solution of Eq. (4.1), then, there are Lagrange multipliers $\lambda^* \geq 0$ and $\mu^* \geq 0$ for $\lambda \geq 0$ and $\mu \geq 0$ that define the following dual function:

$$\phi(\lambda, \mu) = \min_{x \in B(x^*, r)} \left[f(\mathbf{x}) + \lambda^t \mathbf{h}(\mathbf{x}) + \mu^t \mathbf{g}(\mathbf{x}) \right], \qquad (4.15)$$

which is the local minimum of the *Lagrangian* function. The duality theorem proves that the following dual problem:

$$\text{max imize}_{\lambda, \mu} \quad \phi(\lambda, \mu) \qquad (4.16)$$

$$\text{subject to}$$

$$\mu \geq 0$$

has a local solution in $\lambda^* \geq 0$ and $\mu^* \geq 0$, which means that Eqs. (4.1) and (4.16) are equivalent.

Let us consider the MILP problem reformulated as follows:

$$\text{minimize} \quad cx \qquad (4.17)$$

$$\text{subject to}$$

$$Ax \leq b$$

$$Dx \leq e$$

$$x \geq 0, \in N$$

Assuming that the first group of constraints are the ones that make the problem difficult to solve, and considering that λ is a vector of positive multipliers, the problem can be represented as

$$\text{minimize} \quad cx + \lambda(b - Ax) \qquad (4.18)$$

subject to

$$Dx \leq e$$

$$x \geq 0, \in N$$

This modified problem rather than the original one is easier to solve. If the constraints added into the objective function are satisfied, then the objective function benefits, otherwise is penalized. If λ is too big, the constraint is satisfied but far from the limit, then, the solution will not be good. On the other hand, if λ is too small, the constraint may not be satisfied. Iterative methods are developed to find λ values that satisfy the constraint and give the closest value of the objective function with respect to the original unconstrained problem. To update the multipliers, the derivative of the objective function is calculated:

$$L(\lambda) = \text{minimize} \quad cx + \lambda(b - Ax) \qquad (4.19)$$

$$\nabla L(\lambda) = Ax - b, \qquad (4.20)$$

then, the multiplier can be updated in the direction of this derivative. One simple way is to update the multipliers as follows:

$$\lambda_{\text{new}} = \lambda_{\text{old}} + \alpha(Ax - b) \qquad (4.21)$$

Parameter α allows to tune the evolution of the updates. A small α will find the solution slowly while a big α will approach the solution faster but it may oscillate. There are several ways to update the parameter α.

4.4.3.1 *Example*   Let us consider the following maximization problem:

$$\text{maximize} \quad (16x_1 + 10x_2 + 4x_4)$$

subject to

$$8x_1 + 2x_2 + x_3 + 4x_4 \leq 10$$

$$x_1 + x_2 \leq 1$$

$$x_3 + x_4 \leq 1$$

$$x_1, x_2 \geq 0, \in N$$

By relaxing the first row, the problem is formulated as follows:

$$\text{maximize} \quad 16x_1 + 10x_2 + 4x_4 - \lambda(10 - 8x_1 - 2x_2 - x_3 - 4x_4)$$

subject to

$$x_1 + x_2 \leq 1$$
$$x_3 + x_4 \leq 1$$
$$x_1, x_2 \geq 0, \in N$$

Considering an initial $\lambda = 0$ and parameter $\alpha = 1$, in addition to considering that α is reduced by a fraction of third, Table 4.5 illustrates the evolution of the method for the first four iterations, giving the optimal result $x = (1, 0, 0, 1)$.

Table 4.5   LR Example – Iteration Results

Iteration	λ	α	x_1	x_2	x_3	x_4	z
1	0	1	1	0	0	1	20
2	2	$\frac{1}{3}$	0	1	0	0	6
3	0	$\frac{1}{6}$	1	0	0	1	20
4	0.222	$\frac{1}{9}$	1	0	0	1	18

4.4.3.2 Lagrange Relaxation Applied to UC Problems

By inspecting the general formulation of the UC problem, it can be observed that the objective function can be separated by generating units. It can also be observed that some constraints couple the units per period. Therefore, if the constraints that couple the problem can be relaxed, the other part of the problem can be decomposed into independent smaller subproblems, one per generating unit. Lagrange Relaxation methodology takes advantage of this decentralized approach which defines one Lagrange multiplier per constraint that couples the problem and adds a penalty term to the original objective function. For example, considering the energy balance constraint equation, the *Lagrangian* function is

$$\pounds(p, u, \lambda) = C(p_{g,t}, u_{g,t}) + \sum_{t=1}^{T} \lambda^t \left(P_{d,t} - \sum_{g=1}^{M} p_{g,t} \times u_{g,t} \right) \tag{4.22}$$

Lagrange relaxation solves the UC problem by relaxing the coupling constraint through a dual optimization, thus maximizing the *Lagrangian* with respect to the multipliers while it is minimized with respect to the other variables:

$$\max_{\lambda^t} \left\{ \min_{p_{g,t}, u_{g,t}} [\pounds(p, u, \lambda)] \right\} \tag{4.23}$$

This calculation is done in two basic steps:

- Calculate λ^t to maximize

$$\min_{p_{g,t}, u_{g,t}} [\pounds(p, u, \lambda)]$$

- Assuming that the calculated multipliers in the previous step are fixed, minimize the *Lagrangian* adjusting $p_{g,t}$ and $u_{g,t}$ variables.

The second step is calculated using the following equation:

$$\pounds(p, u, \lambda) = \sum_{g=1}^{M} \left[\sum_{t=1}^{T} \left\{ C_i(p_{g,t}) \times u_{g,t} - \lambda^t \times p_{g,t} \times u_{g,t} \right\} \right]$$

since $\sum_{t=1}^{T} (\lambda^t \times P_D^t)$ is constant, and the Lagrangian is separable, the minimum can be calculated by minimizing the generation unit over the time frame:

$$\min_{p_{g,t}, u_i^t} [\pounds(p, u, \lambda)] = \sum_{i=1}^{M} \min_{p_{g,t}, U_i^t} \left[\sum_{t=1}^{T} \left\{ C_i(p_{g,t}) \times u_{g,t} - \lambda^t \times p_{g,t} \times u_{g,t} \right\} \right]$$

subject to the constraints described by Eqs. (4.10)–(4.12). The minimum value per unit can be calculated using progressive dynamic (PD) programming:

$$\min_{p_{g,t}, u_{g,t}} \left[\sum_{t=1}^{T} \left\{ C_i(p_{g,t}) \times u_{g,t} - \lambda^t \times p_{g,t} \times u_{g,t} \right\} \right]$$

subject to

$$u_{g,t} p_g^{\text{Min}} \leq p_{g,t} \leq u_{g,t} p_g^{\text{Max}} \quad \text{for} \quad t = 1, ..., T$$

If the unit g is out of service, then the objective function is zero, otherwise:

$$\min \left[\sum_{t=1}^{T} \left\{ C_i(p_{g,t}) - \lambda^t \times p_{g,t} \right\} \right]$$

subject to the constraints described by Eqs. (4.10)–(4.12).

Since the dynamic programming problem is per generator, the dimensioning problem that affects the PD is not a problem in this case. The problem is an iterative problem that involves two different levels: the main level and the decentralized level.

The dual problem is solved at the main level where the constraint costs or Lagrange multipliers are calculated. The constraint costs are added to the objective function of each subproblem. At the second level, the subproblems are solved. As a result, the multiplier coefficients are adjusted.

It is important to remark that the dual function

$$\phi(\lambda) = \min_{p_{g,t}, u_{g,t}} \ [\pounds(p, u, \lambda)]$$

subject to

$$u_{g,t} p_g^{\text{Min}} \leq p_{g,t} \leq u_{g,t} p_g^{\text{Max}} \quad \text{for} \quad t = 1, ..., T$$

is not differentiable, then the UC problem is not convex due to the integer variables $u_{g,t}$, which is the reason for the existence of the difference between the objective values of the primitive problem and the dual problem:

$$C\left(p_{g,t*}, u_g^{t*}\right) - \phi(\lambda^*)$$

This value is known as *duality gap* which does not depend on the size of the problem, and it satisfies:

$$C\left(p_{g,t}^*, u_g^{t*}\right) \geq \phi(\lambda)$$

Therefore, a quasi optimal solution is expected using Lagrange relaxation method.

4.4.4 Mixed Integer Programming Method

In general, to implement a UC problem using MIP method, the objective function as well as the constraints are linearized. After linearization, the problem becomes an MILP problem. In that case, the *branch and cut* method can be used. The inequality constraints force the problem to include additional integer variables which make the MIP problem hard to solve (NP-hard). One of the methods that is widely used to solve MIP formulations is known as *branch and bound* (B&B). The B&B method solves the MIP problem by solving Linear Programming problems sequentially.

4.4.4.1 Branch and Bound   The B&B method solves the MILP problem by implementing a sequential LP problem which are solved by relaxing the integer condition from the constraints, and including additional constraints.

The number of additional constraints increases as the method evolves through the iterations. Then, it divides the feasible regions into subsets of feasible regions.

Initially, minimum and maximum bounds are set. After that, the strategy is to diminish the upper bound and increase the lower bound. The gap between the bounds establishes the proximity to the optimal solution. The method can complete the calculation due to three possible outcomes: (1) the problem is infeasible, (2) it satisfies the integer conditions, or (3) the lower bound is over the current upper bound. In addition, a particular branch is disabled if one of theses three possibilities is present.

Practically, an MILP problem can be solved by using the B&B methodology by following these steps:

1. **Initialization**: Set the upper and lower bounds. Solve the MILP problem by relaxing the integer constraints. If the problem is infeasible, then there is no possible solution. If the solution is integer, then the optimal solution is achieved. Otherwise, update the lower bound with the calculated solution.

2. **Bifurcation**: Using one of the integer variables that the solution gives as a noninteger, two problems are generated. Let us consider that the variable gets the value $a.b$, where b is the decimal fraction and a is the integer value. Therefore, the problem has two ways to proceed, one with the variable set to a, and another one with the variable set to $(a + 1)$. Both problems can be solved sequentially.

3. **Solution**: Solve the next problem in the sequential list.

4. **Bounds Update**: If the current solution is integer, and the objective function value is less than the upper bound, this solution is saved as an optimal candidate. If the solution is between the bounds, the lower bound is updated with the solution value and another bifurcation is performed.

5. **Cuts**: There are several reasons to avoid a branch: integer solution, the solution is bigger than the upper bound, or an infeasible solution.

6. **Optimality**: If the processing list is not empty, return to step 3, otherwise stop the algorithm and check the list of optimal candidates. If the list is empty, then the problem is infeasible.

Steps 1 and 6 give the results of the algorithm. Any non-integer variable is a candidate for bifurcation, and its selection requires knowledge of the problem that needs a solution.

In order to advance into the subproblems, there are different strategies: (1) breadth-first search or (2) depth-first search. The depth-first search has the advantage of determining faster the upper and lower bounds, an infeasible problem and branch cancellations. The breadth-first search generates initial similar paths that can be exploited during the search. Graphically, the B&B method can be described as illustrated in Figure 4.1.

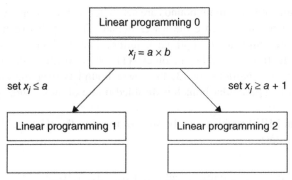

Figure 4.1 Branch and Bound Search.

4.4.4.1.1 Example

$$\text{maximize} \quad z = -3x_1 - 5x_2$$

subject to

$$x_1 + x_2 \leq 8$$
$$2x_1 + 7x_2 \leq 25$$
$$x_1, x_2 \geq 0, \in N$$

Once the mathematical model is defined, the main task is to model the problem to be included into the computational solver. There are several proposals to include them, some of which are reproduced here for an illustrative purpose. Figure 4.2 illustrates the evolution of the B&B methodology applied to this example.

4.4.4.1.2 Production Cost
The formulation shown here is to model a convex model:

$$Cp_{g,t} = u_{g,t}\, c_g + \sum_{b=1}^{B} F_{b,g}\, \delta_{b,g,t} \qquad \forall g, \forall t \qquad (4.24)$$

$$p_{g,t} = u_{g,t}\, \underline{P}_g + \sum_{b=1}^{B} \delta_{b,g,t} \qquad \forall g, \forall t \qquad (4.25)$$

$$(\mathrm{Tr}_{1,g} - \underline{P}_g) \leq \delta_{1,g,t} \qquad \forall g, \forall t \qquad (4.26)$$

$$\delta_{1,g,t} \leq (\mathrm{Tr}_{1,g} - \underline{P}_g) u_{g,t} \qquad \forall g, \forall t \qquad (4.27)$$

$$(\mathrm{Tr}_{b,g} - \mathrm{Tr}_{b-1,g}) \leq \delta_{b,g,t} \qquad \forall g, \forall t, b = 2, ..., B-1 \qquad (4.28)$$

$$\delta_{b,g,t} \leq (\mathrm{Tr}_{b,g} - \mathrm{Tr}_{b-1,g}) \qquad \forall g, \forall t, b = 2, ..., B-1 \qquad (4.29)$$

$$\delta_{B,g,t} \geq 0 \qquad \forall g, \forall t \qquad (4.30)$$

$$\delta_{B,g,t} \leq (\overline{P}_g - \mathrm{Tr}_{B-1,g}) \qquad \forall g, \forall t, \qquad (4.31)$$

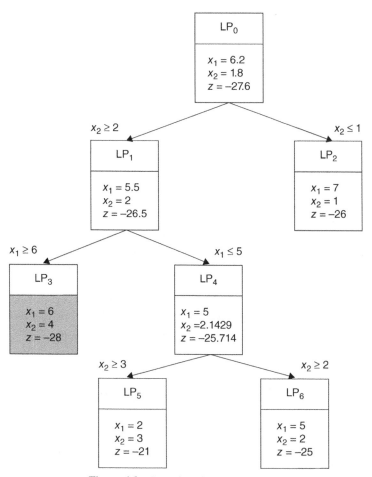

Figure 4.2 Branch and Bound Example.

where $u_{g,t}$ is a binary variable that represents the state of the unit g at time t; c_g is the unit g fixed cost; B is the number of blocks used by the PWL approximation; $F_{b,g}$ is the slope for each power segment b unit g; $\delta_{b,g,t}$ is the MW set for unit g, period t and block b; $p_{g,t}$ is the output MW for unit g period t; $\underline{P}_g$ is the minimum power limit for unit g; $Tr_{b,g}$ are the minimum and maximum power limits within block b unit g; $j_{b,g,t}$ is the binary variable that activate the power block b unit g time t.

4.4.4.1.3 Constraints Modeling   Some of the global and local constraints related to the UC thermal problem are described here to illustrate the complexity associated with the modeling issues in formulating an MILP-based UC problem.

4.4.4.1.4 Energy Balance   In general, it is not difficult to model the global constraints in an MILP UC formulation. Mathematically, the energy balance can be modeled as follows:

$$\sum_{g \in G} P_{g,t} = D_t \qquad \forall t, \qquad (4.32)$$

where D_t is the system demand at period t.

4.4.4.1.5 Local Restrictions   For a thermal plants-based UC, the set of local restrictions usually consist of restrictions on minimum service times, generation unit capacity limits, transition ramps, and logic or binary relationships.

Minimum On/Off Service Times   Constraints related to minimum service times model the operation limitations related to thermal cycles and lifetime of the generation units. There are two constraints: *minimum on service* which is the minimum time that the unit must remain on once it starts running and *minimum off service* time which is the time the unit must remain off once it stops operating.

One way to represent them comprises of six restriction equations that take into account three different simulation periods; the initial period, final period, and intermediate periods. The formulation links three binary variables: state $u_{g,t}$, start $s_{g,t}$, and stop $h_{g,t}$.

$$\sum_{k=1}^{L_g} (1 - u_{g,k}) = 0 \qquad \forall g, L_g = \min \left[T, u_{g0} \left(UT_g - T_g^{on} \right) \right] \qquad (4.33)$$

$$\sum_{i=k}^{k+UT_g-1} u_{g,i} \geq s_{g,k} UT_g \qquad \forall g, \forall k = L_g + 1, \dots, T - UT_g + 1 \qquad (4.34)$$

$$\sum_{i=k}^{T} (u_{g,i} - s_{g,k}) \geq 0 \qquad \forall g, \forall k = T - UT_g + 2, \dots, T \qquad (4.35)$$

$$\sum_{k=1}^{F_g} u_{g,k} = 0 \qquad \forall g, F_g = \min \left[T, (1 - u_{g0}) \left(DT_g - T_g^{off} \right) \right] \qquad (4.36)$$

$$\sum_{i=k}^{k+DT_g-1} (1 - u_{g,i}) \geq h_{g,k} DT_g \qquad \forall g, k = F_g + 1, \dots, T - DT_g + 1 \qquad (4.37)$$

$$\sum_{i=k}^{T} (1 - u_{g,i} - h_{g,k}) \geq 0 \qquad \forall g, k = T - DT_g + 2, \dots, T, \qquad (4.38)$$

where UT_g and DT_g are the required minimum on and off service times, respectively; $s_{g,k}$ represents the start-up binary variable; $h_{g,k}$ is the stop binary variable; T_g^{on} and T_g^{off} are the on/off service times, respectively, that unit g has been at the initial period, $t = 0$.

4.5   NEW CHALLENGES FOR UC PROBLEM

The current implementation of the unit commitment problem faces several new challenges. For large-scale real systems, the natural system of evolution requires handling an increasing number of generators, sometimes in the order of thousands. Therefore, it becomes necessary to develop computational methods to scale the problem and to handle a huge amount of control variables. Besides, the UC problem needs to incorporate different constraints that are becoming necessary such as constraints related to carbon emissions, renewable energy resources, wind farms, PV installations, and storage devices. These types of models may require modeling constraints as quadratic. In addition, it is necessary to incorporate into the model, the unexpected events associated with changes in system interchanges, generation failures, or device outages; these events are associated with discrete disturbances.

Moreover, the increasing penetration of variable generation resources and price-sensitive demands creates new challenges for UC methodologies. Therefore, future solutions to the unit commitment problem should be addressing in effectively dealing with the variability and uncertainty in the problem formulation. For example, to allow the volatile generation or load to be part of the UC solution, it is necessary to improve the accuracy of the forecast tools; model and implement the stochastic UC models to capture the uncertain behavior; and reduce costs or to perform multiple UC over the predetermined horizon, instead of one-shot solution. There are new methodologies that need to be evaluated for UC applications such as stochastic optimization, probabilistic constraints, or robust optimization.

CHAPTER END PROBLEMS

4.1   Consider a power system with the three generators described in Table 4.1 and a fixed demand of 500 MW during 2 hours. The cost functions have the following form:

$$C(P_G) = a + bP_G + cP_G^2$$

Calculate the unit commitment problem using LR method. The initial state is Off for all the units, and both minimum time on and off are set to 1 hour. Set the initial value of the multipliers to 1, no reserve is needed.

4.2 Find all the integer solutions of the following problem using the B&B method. Maximize

$$z = -x_1 + 4x_2$$

subject to

$$
\begin{aligned}
-10x_1 + 20x_2 &\leq 22 \\
5x_1 + 10x_2 &\leq 49 \\
x_1 &\leq 5 \\
x_1, x_2 &\geq 0, \in N
\end{aligned}
$$

4.3 Solve the following integer programming problem using B&B cuts: maximize

$$z = 3x_1 + 4x_2$$

subject to

$$
\begin{aligned}
3x_1 - x_2 &\leq 12 \\
3x_1 + 11x_2 &\leq 66 \\
x_1, x_2 &\geq 0, \in N
\end{aligned}
$$

4.4 Calculate the unit commitment problem of a three-generator system with the parameters described in the following table. The total simulation time is 3 hours, the load profile for these hours is 450 MW, 530 MW, and 600 MW. The minimum time On and Off for all generators is 1 hour, and the initial conditions are On, On, and Off, respectively.

Id	Minimum	Maximum	Cost	Start-up Cost
G_1	60	250	9	1.75
G_2	50	300	8.65	5
G_3	25	65	12	0

FURTHER READING

1. Gruhl J, Schweppe F, Ruane M. Unit commitment scheduling of electric power systems. In: Systems Engineering for Power, Status and Prospects—Proceedings of an Engineering Foundation Conference, August 1975.

2. Sheble G, Fahd G. Unit commitment literature synopsis. *IEEE Transactions on Power Systems* 1994;9(1):128–135.

3. Padhy NP. Unit commitment—a bibliographical survey. *IEEE Transactions on Power Systems* 2004;19(2):55–62.

4. Cohen A, Sherkat V. Optimization-based methods for operations scheduling. *Proceedings of the IEEE* 1987;75(12):1574–1591.

5. Wood AJ, Wollenberg BF. *Power Generation, Operation, and Control*, 2nd edition. New York: John Wiley & Sons; 1996.

6. Li T, Shahidehpour M. Price-based unit commitment: a case of Lagrangian relaxation versus mixed integer programming. *IEEE Transactions on Power Systems* 2005;20(4):2015–2025.

7. Ma H, Shahidehpour M. Unit commitment with transmission security and voltage constraints. *IEEE Transactions on Power Systems* 1999;14(2):757–764.

8. Wolsey LA, Nemhauser GL. *Integer and Combinatorial Optimization*, 1st edition. New York: Wiley-Interscience; 1996.

9. Carrión M, Arroyo JM. A computationally efficient mixed-integer linear formulation for the thermal unit commitment problem. *IEEE Transactions on Power Systems* 2006;21(3):1371–1378.

10. Chang CW, Waight JG. A mixed integer linear programming based hydro unit commitment. In: Proceedings of IEEE Power Engineering Society Summer Meeting, July 1999.

11. Kerr RH, Scheidt JL, Fontanna AJ, Wiley JK. Unit commitment. *IEEE Transactions on Power Apparatus and Systems* 1966;PAS-85(5):417–421.

12. Shoults RR, Chang SK, Helmick S, Grady WM. A practical approach to unit commitment, economic dispatch and savings allocation for multiple-area pool operation with import/export constraints. *IEEE Transactions on Power Apparatus and Systems* 1980;PAS-99(2):625–635.

13. Bellman RE, Dreyfus SE. *Applied Dynamic Programming*. Princeton, NJ: Princeton University Press; 1960.

14. Lowery P. Generating unit commitment by dynamic programming. *IEEE Transactions on Power Apparatus and Systems* 1966;PAS-85(5):422–426.

15. Luenberger DG. *Linear and Nonlinear Programming*. Addison-Wesley; 1989.

16. Bazaraa MS, Sherali HD, Shetty CM. *Nonlinear Programming: Theory and Algorithms*, 3rd edition. Hoboken, NJ: John Wiley & Sons; 2006.

17. Merlin A, Sandrin P. A new method for unit commitment at Électricité de France. *IEEE Transactions on Power Apparatus and Systems* 1983;PAS-102(5):1218–1225.

18. Snyder WL, Powell HD, Rayburn JC. Dynamic programming approach to unit commitment. *IEEE Transactions on Power Systems* 1987;2(2):339–348.

19. Hobbs WJ, Hermon G, Warner S, Shelbe GB. An enhanced dynamic programming approach for unit commitment. *IEEE Transactions on Power Systems* 1988;3(3):1201–1205.

20. Arroyo JM, Conejo AJ. Optimal response of a thermal unit to an electricity spot market. *IEEE Transactions on Power Systems* 2000;15(3):1098–1104.

Chapter 5

Power System Economic Dispatch

Economic dispatch is at the heart of economic operation of a power system. In addition to maintaining the system reliability, meeting the forecasted system load at the lowest possible cost is one of the key goals in power system operation. As stated in Chapter 4, economic dispatch is done after obtaining the solution from the unit commitment problem. Therefore, economic dispatch and unit commitment are essential for optimal system operation.

5.1 INTRODUCTION

The economic dispatch (ED) problem can be defined as the problem that determines the power schedule or generation dispatch of committed generating units to satisfy the required load demand at the lowest possible cost while considering the generating unit limits. The concept of economic dispatch is not new as its initial ideas were conceived back in the 1920s in the United States. George Davison published an article in 1922, where he stated that to establish unit operation schedules, the calculation of fuel consumption for the different turbine combinations at various loads must be done. Two other similar works discussed these same issues. Based on these ideas, several methods were developed and implemented to solve the economic dispatch problem since 1930s. Among these initial methods, the most popular method was the equal incremental cost criterion (EICC). Since then, several other methodologies were developed to improve the production cost minimization algorithm.

Electricity Markets: Theories and Applications, First Edition. Jeremy Lin and Fernando H. Magnago.
© 2017 by The Institute of Electrical and Electronics Engineers, Inc. Published 2017 by John Wiley & Sons, Inc.

5.2 GENERATION COST

The economic dispatch problem primarily depends on the generating unit cost function. Therefore, it is important to know the relationship between cost and output power of a generating unit. The cost to generate an amount of energy (in MWh) can vary widely depending on the unit technology. However, in general, the cost versus power relationship can be represented in four different types of curves:

1. Input–output curve
2. Fuel cost curve
3. Heat rate curve
4. Incremental cost curve

5.2.1 Input–Output Curve

The input–output curve can be built experimentally using field data, by varying the fuel consumption and measuring the power output for that consumption. For each unit, there must be both the minimum and the maximum MW values (aka operating limits). Figure 5.1 illustrates a typical input–output curve for a thermal generating unit.

5.2.2 Fuel Cost Curve

The fuel-cost curve for a generator represents a function of the cost of electricity generation with respect to the power output. For most generators, the majority of the generation cost is comprised of the fuel cost. Other components of the generation cost

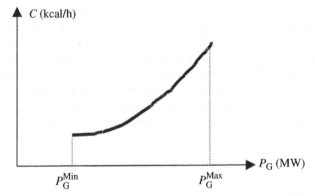

Figure 5.1   Input-Output Curve of a Thermal Generating Unit.

include variable O&M (operation and maintenance), and emission cost. The fuel-cost curve for a generating unit i can be considered as a quadratic polynomial function:

$$C(P_{G_i}) = a_i + b_i P_{G_i} + c_i P_{G_i}^2 \tag{5.1}$$

The coefficients (a_i, b_i, c_i) for unit i can be obtained from the unit design data or from field measurements. The operating limits of unit i can be represented as follows:

$$P_{G_i}^{Min} \leq P_{G_i} \leq P_{G_i}^{Max} \tag{5.2}$$

The key problem of the curve shown in Eq. (5.1), is that the quadratic function makes the optimization problem nonlinear. However, to overcome this nonlinearity problem, the quadratic function can be approximated by a linear or piecewise linear curve with the goal of converting the nonlinear problem into a linear one.

5.2.3 Heat Rate Curve

Using the unit's fuel cost ($/UComb) and the specific heat rate (HR) (kcal/UComb), the hourly fuel consumption can be converted to the hourly production cost ($/h):

$$\left[\frac{\$}{h}\right] = \left[\frac{kcal}{h}\right] \left[\frac{\$}{UComb}\right] \frac{1}{HR} \tag{5.3}$$

The HR can be obtained based on standard measurements using a calorimetry, obtaining two values: (i) *maximum heat rate*, which includes the vaporization heat value and (ii) *minimum heat rate*, which does not include this term. The main difference between the two is that the hydrogen percent is included in the fuel. Figure 5.2 shows a typical heat rate curve for a generator.

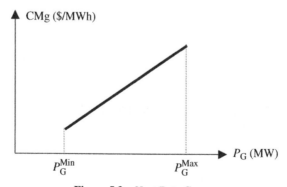

Figure 5.2 Heat Rate Curve.

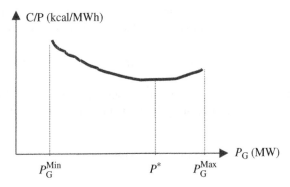

Figure 5.3 Incremental Cost Curve.

5.2.4 Incremental Cost Curve

The incremental cost curve, which is the derivative of the cost curve, is another useful way to represent the cost versus power relationship. If Eq. (5.1) is differentiated with respect to P_{G_i}, the following equation can be obtained:

$$\frac{dC(P_{G_i})}{dP_{G_i}} = b_i + 2c_i P_{G_i} \qquad (5.4)$$

Also, the incremental cost curve represents the incremental change in the input cost when a unit increases or decreases the power generation by one unit. Figure 5.3 represents a typical incremental cost curve.

Finally, one additional important feature of the heat rate is its relation with the unit efficiency. The following equation describes this relationship:

$$\eta = \frac{1}{\dfrac{C}{P_G}} f_{\text{conv}} \qquad (5.5)$$

The conversion factor f_{conv} is due to the following equivalences:

$$1 \text{ kWh} = 860 \text{ kcal} = 3.412 \text{ Btu}$$

5.3 MATHEMATICAL FORMULATION FOR ECONOMIC DISPATCH

Mathematically, the economic dispatch problem can be formulated using the constrained static optimization problem defined by Eq. (4.1). The first-order optimality condition or *Karush–Kuhn–Tucker* (KKT) requirement establishes that if $\mathbf{x}^*$ is an optimal solution, then $\exists \; \lambda_i \in R, i = 1, \ldots, r$ and $\mu_j \geq 0, j = 1, \ldots, p$; such that

$$\nabla \pounds(\mathbf{x}^*) = \mathbf{0}$$
$$\mu_j g_j(\mathbf{x}^*) = 0; \quad j = 1, \ldots, p \tag{5.6}$$

where the function:

$$\pounds = f + \lambda^t \mathbf{h} + \mu^t \mathbf{g}$$

is known as *Lagrange function*.

5.4 ECONOMIC DISPATCH PROBLEM

From the general formulation, the economic dispatch problem can be defined as a non-linear optimization problem as follows:

$$\text{minimize} \quad \left\{ \sum_{i=1}^{M} C_i(P_{G_i}) \right\} \tag{5.7}$$

subject to

$$\sum_{i=1}^{M} P_{G_i} = P_L + \sum_{i=1}^{N} P_{D_i}$$

$$P_{G_i}^{\text{Min}} \leq P_{G_i} \leq P_{G_i}^{\text{Max}}; \quad i = 1, \ldots, M,$$

where M is the number of generating units and N is the number of demand nodes. The objective function is the minimization of the total active power generation cost, represented by the function C in which the key variable is the generating unit active power P_{G_i}. This function is nonlinear. However, it can be simplified by replacing the nonlinear function with a linear piecewise function, making the problem a linear optimization problem. The first constraint represents the power balance equation in which P_{D_i} is the demand at node i and P_L is the total active power loss. The second constraint represents the minimum and maximum operating limits of each generating unit.

In order to establish the first-order optimality condition, the Lagrange function can be defined as

$$\pounds = \sum_{i=1}^{M} C_i(P_{G_i}) + \lambda \left[P_L + \sum_{i=1}^{N} P_{D_i} - \sum_{i=1}^{M} P_{G_i} \right] \tag{5.8}$$

$$+ \sum_{i=1}^{M} \left[\mu_i^{\text{Min}} \left(P_{G_i}^{\text{Min}} - P_{G_i} \right) + \mu_i^{\text{Max}} \left(P_{G_i} - P_{G_i}^{\text{Max}} \right) \right],$$

where λ represents the Lagrange multiplier of the power balance equation, and μ_i^{Min} and μ_i^{Max} represent the corresponding Lagrange multipliers of power limit equations. From this equation, the first-order optimality conditions (KKT) can be derived as follows:

$$\text{For} \quad P_{G_i}^{\text{Min}} < P_{G_i} < P_{G_i}^{\text{Max}}$$

$$\frac{dC_i(P_{G_i})}{dP_{G_i}} + \lambda\left(\frac{\partial P_L}{\partial P_{G_i}} - 1\right) = 0$$

$$\text{if} \quad P_{G_i} = P_{G_i}^{\text{Min}}$$

$$\frac{dC_i(P_{G_i})}{dP_{G_i}} + \lambda\left(\frac{\partial P_L}{\partial P_{G_i}} - 1\right) - \mu_i^{\text{Min}} = 0; \quad \mu_i^{\text{Min}} \geq 0 \tag{5.9}$$

$$\text{if} \quad P_{G_i} = P_{G_i}^{\text{Max}}$$

$$\frac{dC_i(P_{G_i})}{dP_{G_i}} + \lambda\left(\frac{\partial P_L}{\partial P_{G_i}} - 1\right) + \mu_i^{\text{Max}} = 0; \quad \mu_i^{\text{Max}} \geq 0$$

$$\sum_{i=1}^{M} P_{G_i} = P_L + \sum_{i=1}^{N} P_{D_i}$$

5.5 LOSSLESS ECONOMIC DISPATCH FORMULATION

For simplicity, the losses can be initially neglected, then $P_L = 0$. The problem can be considered without the network, where all generators and loads are connected to a single bus bar, as illustrated in Figure 5.4. This approximation allows us to analyze the problem more conveniently.

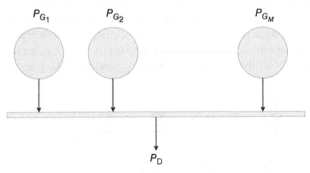

Figure 5.4 Single Bus Dispatch.

The first-order optimality conditions (KKT) are

$$\text{For} \quad P_{G_i}^{\text{Min}} < P_{G_i} < P_{G_i}^{\text{Max}}$$

$$\frac{dC_i(P_{G_i})}{dP_{G_i}} = \lambda$$

$$\text{if} \quad P_{G_i} = P_{G_i}^{\text{Min}}$$

$$\frac{dC_i(P_{G_i})}{dP_{G_i}} - \mu_i^{\text{Min}} = \lambda; \quad \mu_i^{\text{Min}} \geq 0 \qquad (5.10)$$

$$\text{if} \quad P_{G_i} = P_{G_i}^{\text{Max}}$$

$$\frac{dC_i(P_{G_i})}{dP_{G_i}} + \mu_i^{\text{Max}} = \lambda; \quad \mu_i^{\text{Max}} \geq 0$$

$$\sum_{i=1}^{M} P_{G_i} = \sum_{i=1}^{N} P_{D_i}$$

All generating units that are operating within their operating limits will be working at the same marginal cost value. This value, normally represented by the symbol lambda is the Lagrange Multiplier. The units which are more expensive than lambda will operate at their minimum power and the units cheaper than lambda will operate at their maximum power.

$$\mu_i^{\text{Min}} \geq 0 \implies \frac{dC_i}{dP_{G_i}}\left(P_{G_i}^{\text{Min}}\right) = \lambda + \mu_i^{\text{Min}} \geq \lambda$$

$$\mu_i^{\text{Max}} \geq 0 \implies \frac{dC_i}{dP_{G_i}}\left(P_{G_i}^{\text{Max}}\right) = \lambda - \mu_i^{\text{Max}} \leq \lambda$$

As an example, Figure 5.5 shows the dispatch selection for a two-unit system when λ is within the operating cost range of both machines, Figure 5.6 illustrates the case of a cheap unit and Figure 5.7 depicts the case of an expensive unit.

The cost function parameters considered for this example are described in Table 4.1

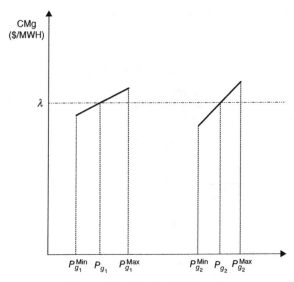

Figure 5.5   Two Units Dispatch.

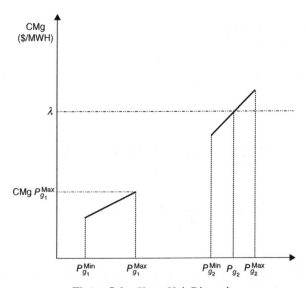

Figure 5.6   Cheap Unit Dispatch.

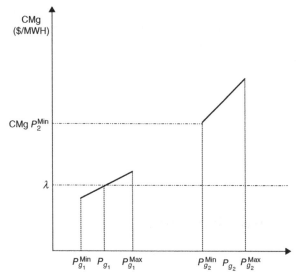

Figure 5.7 Expensive Unit Dispatch.

From KKT conditions,

$$\frac{dC_{G_1}}{dP_{G_1}} = 2227 + 2 \times 9.60\,P_{G_1} = \lambda$$

$$\frac{dC_{G_2}}{dP_{G_2}} = 2373 + 2 \times 14.20\,P_{G_2} = \lambda$$

$$\frac{dC_{G_3}}{dP_{G_3}} = 2500 + 2 \times 18.95\,P_{G_3} = \lambda$$

$$P_{G_1} + P_{G_2} + P_{G_3} = P_{\mathrm{L}}$$

The solution for these equations is $P_{G_1} = 234.7$ MW; $P_{G_2} = 153.6$ MW; $P_{G_3} = 111.7$ MW; $\lambda = 6.734$ kcal/MWh. This solution meets the load constraints and all units are within the limits. Then, all units will operate at the same marginal cost which is the system λ.

5.6 NUMERICAL METHODS FOR ECONOMIC DISPATCH

Several methods have been proposed to solve the ED problem. These methods can be classified into three main groups: the heuristics, the primal, and the penalty and barrier methods.

- **Heuristics**: Heuristic methods take advantage of the structure of the problem. The most well-known heuristic methods are the lambda iteration, base point, and participation factor methods.
- **Primal**: Primal methods are based on the algorithm on the direct search of the optimal solution. Some examples of primal methods are the reduced gradient method or the generalized reduced gradient methods.
- **Penalty and Barrier**: In these methods, the constrained problem is approximated by an unconstrained problem.

To illustrate the different methods, some of the simplest methods are explained next.

5.6.1 Lambda Iteration Method

This method is presented here because of its simplicity. However, it is important to remark that it may present convergence problems. The flow chart illustrated in Figure 5.8 shows the steps of the algorithm.

Considering the three units example presented before, let us consider an initial value of $\lambda^{(1)} = 5000$[kcal/MWh]. Then, the KKT conditions are

$$\frac{dC_{G_1}}{dP_{G_1}} = 2227 + 2 \times 9.60 P_{G_1} = 5000$$

$$\frac{dC_{G_2}}{dP_{G_2}} = 2373 + 2 \times 14.20 P_{G_2} = 5000$$

$$\frac{dC_{G_3}}{dP_{G_3}} = 2500 + 2 \times 18.95 P_{G_3} = 5000$$

which gives the following initial dispatch: $P_{G_1} = 144.43$ MW, $P_{G_2} = 92.5$ MW, and $P_{G_3} = 65.96$ MW. In the first iteration, the limits are relaxed. The initial error is

$$e^{(0)} = 500 - (144.43 + 92.50 + 65.96) = 197.11 \text{ MW}$$

The positive error indicates that more generation is needed, or a larger λ value. Setting now $\lambda^{(1)} = 7500$ [kcal/MWh], and calculating the dispatch again provides: $P_{G_1} = 274.63$ MW, $P_{G_2} = 180.53$ MW, and $P_{G_3} = 131.93$ MW. Then, the updated error is

$$e^{(2)} = 500 - (274.63 + 180.53 + 131.93) = -87.1 \text{ MW}$$

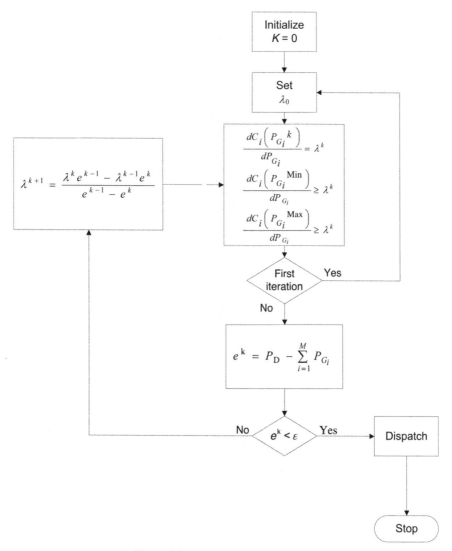

Figure 5.8   Lambda Iteration Method.

Now using interpolation, a new λ is obtained:

$$\lambda^{(2)} = \frac{\lambda^{(1)}e^{(0)} - \lambda^{(0)}e^{(1)}}{e^{(0)} - e^{(1)}} = \frac{7500 \times 197.11 - 5000 \times (-87.1)}{197.11 - (-87.1)}$$

$$\lambda^{(2)} = 6733.85 \ (\text{kcal/MWh})$$

Table 5.1 summarizes the calculation results for this and another iteration.

Table 5.1   Results for Iteration Two and Three

Iteration	G_1	G_2	G_3	Error	λ
2	274.63	180.3	131.93	−87.1	6733.85
3	234.73	153.55	111.71	0.0061	

5.6.2 Newton–Raphson Method

The aim of the ED is to solve Eq. (5.6). Since it is a vector function, the formulation can be done in order to find the deviation that moves the gradient close to zero. Then, the Newton–Raphson method can be used for this propose.

The optimal solution needs to comply with the KKT requirements, in particular:

$$\nabla \pounds(\mathbf{x}^*) = \mathbf{0} \tag{5.11}$$

If the generating unit's limit constraints are also considered, then, the *Lagrangian* function becomes

$$\pounds = \sum_{i=1}^{M} C_i(P_{G_i}) + \lambda \left[\sum_{i=1}^{N} P_{D_i} - \sum_{i=1}^{M} P_{G_i} \right] \tag{5.12}$$

Then,

$$\nabla \pounds(\mathbf{x}) = \begin{bmatrix} \dfrac{\partial \pounds}{\partial P_{G_1}} \\ \vdots \\ \dfrac{\partial \pounds}{\partial P_{G_M}} \\ \dfrac{\partial \pounds}{\partial \lambda} \end{bmatrix} = \begin{bmatrix} \dfrac{dC_1(P_{G_1})}{dP_{G1}} - \lambda \\ \vdots \\ \dfrac{dC_M(P_{G_M})}{dP_{GM}} - \lambda \\ \sum_{i=1}^{N} P_{D_i} - \sum_{i=1}^{M} P_{G_i} \end{bmatrix}, \tag{5.13}$$

where

$$\mathbf{x} = \begin{bmatrix} P_{G1} \\ \vdots \\ P_{G_M} \\ \lambda \end{bmatrix}$$

As a result, the following matrix system can be obtained:

$$\left[\frac{\partial \nabla \pounds(\mathbf{x})}{\partial \mathbf{x}} \right] \Delta \mathbf{x} = -\nabla \pounds(\mathbf{x}) \tag{5.14}$$

And the *Lagrangian Hessian* system becomes

$$
\frac{\partial \nabla \pounds(\mathbf{x})}{\partial \mathbf{x}} =
\begin{bmatrix}
\dfrac{d^2\pounds}{dP_{G_1}^2} & \dfrac{d^2\pounds}{dP_{G_1}dP_{G2}} & \cdots & \dfrac{d^2\pounds}{dP_{G_1}dP_{G_M}} & \dfrac{d^2\pounds}{dP_{G_1}d\lambda} \\[2ex]
\dfrac{d^2\pounds}{dP_{G_2}dP_{G_1}} & \dfrac{d^2\pounds}{dP_{G_2}^2} & \cdots & \dfrac{d^2\pounds}{dP_{G_2}dP_{G_M}} & \dfrac{d^2\pounds}{dP_{G_2}d\lambda} \\[2ex]
\vdots & \vdots & \ddots & \vdots & \vdots \\[2ex]
\dfrac{d^2\pounds}{dP_{G_M}dP_{G_1}} & \dfrac{d^2\pounds}{dP_{G_M}dP_{G_2}} & \cdots & \dfrac{d^2\pounds}{dP_{G_M}^2} & \dfrac{d^2\pounds}{dP_{G_M}d\lambda} \\[2ex]
\dfrac{d^2\pounds}{d\lambda dP_{G_1}} & \dfrac{d^2\pounds}{d\lambda dP_{G_2}} & \cdots & \dfrac{d^2\pounds}{d\lambda dP_{G_M}} & \dfrac{d^2\pounds}{d\lambda^2}
\end{bmatrix}
\tag{5.15}
$$

The results of the second derivative of the *Lagrangian* are

$$
\frac{\partial \nabla \pounds(\mathbf{x})}{\partial \mathbf{x}} =
\begin{bmatrix}
\dfrac{d^2C_1(P_{G_1})}{dP_{G_1}^2} & 0 & \cdots & 0 & -1 \\[2ex]
0 & \dfrac{d^2C_2(P_{G2})}{dP_{G_2}^2} & \cdots & 0 & -1 \\[2ex]
\vdots & \vdots & \ddots & \vdots & \vdots \\[2ex]
0 & 0 & \cdots & \dfrac{d^2C_M(P_{G_M})}{dP_{G_M}^2} & -1 \\[2ex]
-1 & -1 & \cdots & -1 & 0
\end{bmatrix}
\tag{5.16}
$$

The solution at iteration k is

$$
\left[\frac{\partial \nabla \pounds(\mathbf{x}^{(k)})}{\partial \mathbf{x}}\right] \Delta \mathbf{x}^{(k)} = -\nabla \pounds(\mathbf{x}^{(k)})
\tag{5.17}
$$

Then, the new operating points can be obtained:

$$
\mathbf{x}^{(k+1)} = \mathbf{x}^{(k)} + \Delta \mathbf{x}^{(k)}
\tag{5.18}
$$

The flow chart described in Figure 5.9 summarizes the method steps.
For the three units example, let us first consider an initial feasible dispatch:

$$
\mathbf{x}^{(0)} =
\begin{bmatrix}
P_{G_1}^{(0)} \\
P_{G_2}^{(0)} \\
P_{G_3}^{(0)} \\
\lambda^{(0)}
\end{bmatrix}
=
\begin{bmatrix}
200 & \text{MW} \\
180 & \text{MW} \\
120 & \text{MW} \\
5000 & \text{kcal/MWh}
\end{bmatrix}
$$

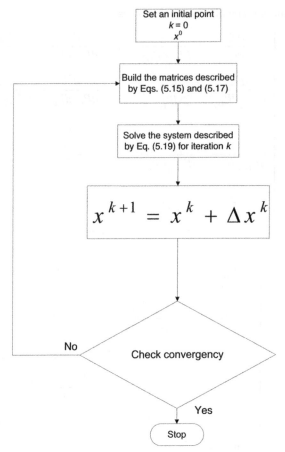

Figure 5.9   Newton–Raphson Method.

Then, the Lagrangian gradient is

$$
\nabla \pounds(\mathbf{x}^{(0)}) =
\begin{bmatrix}
\dfrac{dC_{G_1}(P_{G_1}^{(0)})}{dP_{G_1}} - \lambda^{(0)} \\[2ex]
\dfrac{dC_{G_2}(P_{G_2}^{(0)})}{dP_{G_2}} - \lambda^{(0)} \\[2ex]
\dfrac{dC_{G_3}(P_{G_3}^{(0)})}{dP_{G_3}} - \lambda^{(0)} \\[2ex]
500 - (P_{G_1}^{(0)} + P_{G_2}^{(0)} + P_{G_3}^{(0)})
\end{bmatrix}
=
\begin{bmatrix}
1.067 \\
2.485 \\
2.048 \\
0
\end{bmatrix}
$$

And the Hessian matrix is

$$\frac{\partial \nabla \pounds(\mathbf{x}^{(0)})}{\partial \mathbf{x}} = \begin{bmatrix} \dfrac{d^2 C_{G_1}(P_{G_1}^{(0)})}{dP_{G_1}^2} & 0 & 0 & -1 \\[2ex] 0 & \dfrac{d^2 C_{G_2}(P_{G_2}^{(0)})}{dP_{G_2}^2} & 0 & -1 \\[2ex] 0 & 0 & \dfrac{d^2 C_{G_3}(P_{G_3}^{(0)})}{dP_{G_3}^2} & -1 \\[2ex] -1 & -1 & -1 & 0 \end{bmatrix}$$

$$\frac{\partial \nabla \pounds(\mathbf{x}^{(0)})}{\partial \mathbf{x}} = \begin{bmatrix} 19.20 & 0 & 0 & -1 \\ 0 & 28.40 & 0 & -1 \\ 0 & 0 & 37.90 & -1 \\ -1 & -1 & -1 & 0 \end{bmatrix}$$

Solving the linear system:

$$\left[\frac{\partial \nabla \pounds(\mathbf{x}^{(0)})}{\partial \mathbf{x}} \right] \Delta \mathbf{x}^{(0)} = -\nabla \pounds(\mathbf{x}^{(0)})$$

the variables updates are

$$\Delta \mathbf{x}^{(0)} = \begin{bmatrix} \Delta P_{G_1}^{(0)} \\[1ex] \Delta P_{G_2}^{(0)} \\[1ex] \Delta P_{G_3}^{(0)} \\[1ex] \Delta \lambda^{(0)} \end{bmatrix} = \begin{bmatrix} 34.7 & \text{MW} \\ -26.4 & \text{MW} \\ -8.3 & \text{MW} \\ 1733.9 & \text{kcal/MWh} \end{bmatrix}$$

And the new operation point is

$$\begin{bmatrix} P_{G_1}^{(1)} \\[1ex] P_{G_2}^{(1)} \\[1ex] P_{G_3}^{(1)} \\[1ex] \lambda^{(1)} \end{bmatrix} = \begin{bmatrix} P_{G_1}^{(0)} \\[1ex] P_{G_2}^{(0)} \\[1ex] P_{G_3}^{(0)} \\[1ex] \lambda^{(0)} \end{bmatrix} + \begin{bmatrix} \Delta P_{G_1}^{(0)} \\[1ex] \Delta P_{G_2}^{(0)} \\[1ex] \Delta P_{G_3}^{(0)} \\[1ex] \Delta \lambda^{(0)} \end{bmatrix}$$

$$
\begin{bmatrix} P^{(1)}_{G_1} \\ P^{(1)}_{G_2} \\ P^{(1)}_{G_3} \\ \lambda^{(1)} \end{bmatrix} = \begin{bmatrix} 200 & \text{MW} \\ 180 & \text{MW} \\ 120 & \text{MW} \\ 5000 & \text{kcal/MWh} \end{bmatrix} + \begin{bmatrix} 34.7 & \text{MW} \\ -26.4 & \text{MW} \\ -8.3 & \text{MW} \\ 1733.9 & \text{kcal/MWh} \end{bmatrix}
$$

$$
\begin{bmatrix} P^{(1)}_{G_1} \\ P^{(1)}_{G_2} \\ P^{(1)}_{G_3} \\ \lambda^{(1)} \end{bmatrix} = \begin{bmatrix} 234.7 & \text{MW} \\ 153.6 & \text{MW} \\ 111.7 & \text{MW} \\ 6733.9 & \text{kcal/MWh} \end{bmatrix}
$$

This final dispatch is feasible.

5.6.3 Reduced Gradient Methods

This method was initially developed to solve nonlinear problems with linear restrictions. However, it was later extended to solve nonlinear problems with nonlinear constraints. Let us consider the following objective function:

$$
C(\mathbf{P}) = \sum_{i=1}^{M} C_i(P_{G_i}), \tag{5.19}
$$

where

$$
\mathbf{P} = \begin{bmatrix} P_1 \\ P_2 \\ \vdots \\ P_M \end{bmatrix} \tag{5.20}
$$

If the Taylor series is truncated by considering only the first-order term, the following approximate expression can be obtained:

$$
\Delta C(\mathbf{P}) = \nabla C(\mathbf{P})^t \Delta \mathbf{P}, \tag{5.21}
$$

where $\nabla C(\mathbf{P})$ is the gradient of the objective function valuated on $\mathbf{P}$. Always, one generator is a dependent variable and can be removed from the formulation. For simplicity, if it is a fixed demand, then, the power variation of one generator is compensated by the power variation of other generators in the opposite direction. In other words, the sum of the power variation of all generators in the system must be zero. This can be written as

$$\sum_{i=1}^{M} P_{G_i} = \sum_{i=1}^{N} P_{D_i} \Longrightarrow \sum_{i=1}^{M} \Delta P_{G_i} = 0 \tag{5.22}$$

Consider the following equation:

$$\Delta P_{G_k} = -\sum_{i \neq k} \Delta P_{G_i} \tag{5.23}$$

It is possible to rewrite the change in the objective function by its reduced gradient:

$$\Delta C(\mathbf{P}) = \sum_{i \neq k} \left[\frac{dC_i(P_{G_i})}{dP_{G_i}} - \frac{dC_k(P_{G_k})}{dP_{G_k}} \right] \Delta P_{G_i} \tag{5.24}$$

Based on this formulation, the algorithm steps can be summarized with the flowchart illustrated in Figure 5.10:

Using the three units data, the initial feasible dispatch for a 500 MW load is

$$P_{G_1}^{(0)} = 200 \text{ MW}$$

$$P_{G_2}^{(0)} = 180 \text{ MW}$$

$$P_{G_3}^{(0)} = 120 \text{ MW}$$

The total cost is $C(\mathbf{P})^{(0)} = 2335605$ kcal/h Selecting G_3 as the dependent unit, the reduced gradient for the total cost formulation is

$$\frac{dC_{G_1}(P_{G_1}^{(0)})}{dP_{G_1}} = 2227 + 2 \times 9.60 P_{G_1}^{(0)} = 6067$$

$$\frac{dC_{G_2}(P_{G_2}^{(0)})}{dP_{G_2}} = 2373 + 2 \times 14.20 P_{G_2}^{(0)} = 7485$$

$$\frac{dC_{G_3}(P_{G_3}^{(0)})}{dP_{G_3}} = 2500 + 2 \times 18.95 P_{Pilar}^{(0)} = 7048$$

Then,

$$\frac{dC_{G_1}(P_{G_1}^{(0)})}{dP_{G_1}} - \frac{dC_{G_3}(P_{G_3}^{(0)})}{dP_{G_3}} = -981$$

$$\frac{dC_{G_2}(P_{G_2}^{(0)})}{dP_{G_2}} - \frac{dC_{G_3}(P_{G_3}^{(0)})}{dP_{G_3}} = 437$$

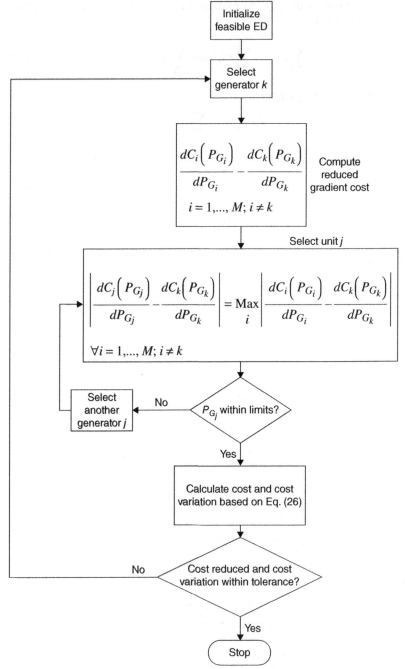

Figure 5.10 Reduced Gradient Method.

The cost variation is

$$\Delta C(\mathbf{P}) = (-981)\Delta P_{G_1} + (437)\Delta P_{G_2}$$

The larger coefficient is related to G_1, and is negative, then the generation for G_1 is incremented and that for G_3 is decreased by the same value:

$$P_{G_1}^{(1)} = 220 \text{ MW}$$

$$P_{G_2}^{(1)} = 180 \text{ MW}$$

$$P_{G_3}^{(1)} = 100 \text{ MW}$$

This is a feasible dispatch and the total cost is $C(P)^{(1)} = 2327405$ kcal/h $< C(P)^{(0)}$. The cost variation between the two iterations is $C(P)^{(1)} - C(P)^{(0)} = 8200$.

Next iteration is

$$\frac{dC_{G_1}(P_{G_1}^{(1)})}{dP_{G_1}} - \frac{dC_{G_3}(P_{G_3}^{(1)})}{dP_{G_3}} = 161$$

$$\frac{dC_{G_2}(P_{G_2}^{(1)})}{dP_{G_2}} - \frac{dC_{G_3}(P_{G_3}^{(1)})}{dP_{G_3}} = 1195$$

Now, if the biggest coefficient corresponds to G_2, and is positive, then the generation of G_2 is decreased and the same amount is increased for the dispatch of unit G_3:

$$P_{G_1}^{(2)} = 220 \text{ MW}$$

$$P_{G_2}^{(2)} = 160 \text{ MW}$$

$$P_{G_3}^{(2)} = 120 \text{ MW}$$

Again, this is a feasible dispatch, the total cost is $C(P)^{(2)} = 2316765$ kcal/h $<$ $C(P)^{(1)}$, and the cost difference is $\{C(P)^{(2)} - C(P)^{(1)}\} = 10640$. The process continues until this cost variation is less than a predetermined threshold.

5.7 INCLUSION OF TRANSMISSION LOSSES

In the previous analysis, the influence of losses is neglected because the system is modeled as a single node system. The impact of network losses should be included for economic dispatch calculation. And, the system can be represented as shown in

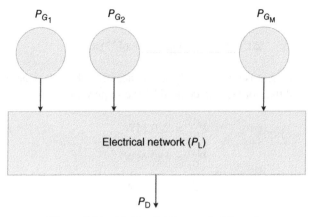

Figure 5.11   Multinodal Economic Dispatch.

Figure 5.11. The losses are functions of the generation dispatch. If the generation plant is located closer to the load, then, the impact of losses can be reduced. However, the generating plants dispersedly located at different places interconnected by long transmission lines increase the effect of losses. Therefore, generators not only need to satisfy system load but also the network losses.

Let us define a penalization factor due to load for each generation node as follows:

$$fp_i = \frac{1}{1 - \frac{\partial P_L}{\partial P_{G_i}}}, \tag{5.25}$$

where $\frac{\partial P_L}{\partial P_{G_i}}$ is known as the incremental loss at node i. If an increase of generation at node i causes an increase on total losses, then, the incremental loss at node i is positive:

$$\frac{\partial P_L}{\partial P_{G_i}} > 0$$

and the penalization factor is

$$fp_i > 1$$

On the other hand, if an increase of generation at node i causes a decrease on the total system losses, then, the incremental loss at node i is negative:

$$\frac{\partial P_L}{\partial P_{Gi}} < 0$$

and the penalization factor is

$$fp_i < 1$$

Therefore, the optimal economic dispatch is defined when all units that operate within their limits operate at the same penalized marginal cost, which becomes the system lambda.

$$fp_i \frac{dC_i(P_{G_i})}{dP_{G_i}} = \lambda \qquad (5.26)$$

Similarly, if the unit is expensive, its penalized marginal cost becomes greater than or equal to the system lambda and the unit will be dispatched at its minimum value.

$$\mu_i^{Min} \geq 0; \quad fp_i \geq 0 \Longrightarrow fp_i \frac{dC_i}{dP_{G_i}} \left(P_{G_i}^{Min}\right) = \lambda + \frac{\mu_i^{Min}}{fp_i} \geq \lambda \qquad (5.27)$$

On the contrary, if the unit is cheaper, its penalized marginal cost becomes less or equal than the system lambda and the unit is dispatched at its maximum value.

$$\mu_i^{Max} \geq 0; \quad fp_i \geq 0 \Longrightarrow fp_i \frac{dC_i}{dP_{G_i}} \left(P_{G_i}^{Max}\right) = \lambda - \frac{\mu_i^{Max}}{fp_i} \leq \lambda \qquad (5.28)$$

The effect of losses on the generation dispatch is illustrated in Figure 5.12.

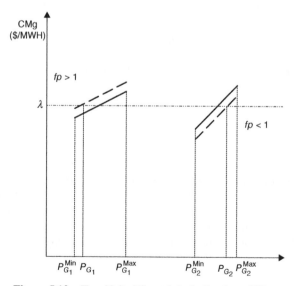

Figure 5.12 Two Units Dispatch Including Loss Effect.

5.7.1 Total and Incremental Loss Calculation

Two main methodologies are used to calculate the incremental and total losses as well as the corresponding penalization factors:

1. Loss formulae (Matrix B)
2. Power flow calculation

The use of loss formulae only allows an approximate calculation because its development assumes a set of hypothesis. For example, the matrix B formula is

$$P_L = \mathbf{P}^t \mathbf{B} \mathbf{P} + \mathbf{P}^t \mathbf{B}_0 + B_{00} P_L = \sum_{i=1}^{M} \sum_{j=1}^{M} P_{G_i} B_{ij} P_{G_j} + \sum_{i=1}^{M} B_{0i} P_{G_i} + B_{00}, \quad (5.29)$$

where $\mathbf{P}$ is the generation vector, $\mathbf{B}$ the loss matrix, $\mathbf{B}_0$ is the loss vector and B_{00} is a constant. The assumptions considered are

1. Constant voltages
2. Individual loads are linearly related to total load:

$$I_i = \alpha_i I_D + I_i^0 \quad (5.30)$$

3. Reactive power injection is a linear function of active power injection:

$$Q_i = \beta_i P_{G_i} + Q_i^0 \quad (5.31)$$

Then, the marginal losses can be represented as follows:

$$\frac{\partial P_L}{\partial P_{G_i}} = 2 \sum_{j=1}^{M} B_{ij} P_{G_j} - B_{00} \quad (5.32)$$

A basic algorithm to solve the multinodal ED using matrix B, considering the losses can be performed as shown in the flowchart of Figure 5.13.

The coordination equations can be solved using the techniques already described in the previous section. Based on this formulation, several authors proposed different alternatives to calculate the coefficients of matrix B.

In order to illustrate the calculation methods, one methodology based on the power flow formulation is described next.

$$P_{G_i}^{new} = P_{G_i}^{old} + \Delta P_{G_i} \quad (5.33)$$

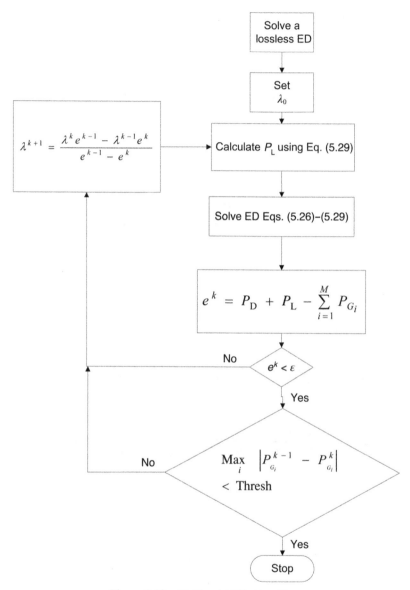

Figure 5.13 Multinodal ED Algorithm.

Let us consider a constant load. To compensate the power variation of one generation, a change on the reference node is produced:

$$P_{\text{Slack}}^{\text{new}} = P_{\text{Slack}}^{\text{old}} + \Delta P_{\text{Slack}} \tag{5.34}$$

The power flows may change due to these changes, and accordingly the total losses can change:

$$\Delta P_{\text{Slack}} + \Delta P_{G_i} = \Delta P_L, \tag{5.35}$$

defining,

$$\beta_i = \frac{-\Delta P_{\text{Slack}}}{\Delta P_{G_i}} = \frac{\Delta P_{G_i} - \Delta P_L}{\Delta P_{G_i}} = 1 - \frac{\partial P_L}{\partial P_{G_i}},$$

then,

$$fp_i = \frac{1}{\beta_i} \tag{5.36}$$

Using a power flow:

$$\Delta P_{\text{Slack}} = \sum_{i=1}^{M} \frac{\partial P_{\text{Slack}}}{\partial \theta_i} \frac{\partial \theta_i}{\partial P_{G_i}} \Delta P_{G_i} + \sum_{i=1}^{M} \frac{\partial P_{\text{Slack}}}{\partial V_i} \frac{\partial V_i}{\partial P_{G_i}} \Delta P_{G_i} \tag{5.37}$$

$$\Delta P_{\text{Slack}} = \sum_{i=1}^{M} \frac{\partial P_{\text{Slack}}}{\partial \theta_i} \frac{\partial \theta_i}{\partial Q_{G_i}} \Delta Q_{G_i} + \sum_{i=1}^{M} \frac{\partial P_{\text{Slack}}}{\partial V_i} \frac{\partial V_i}{\partial Q_{G_i}} \Delta Q_{G_i} \tag{5.38}$$

In matrix form,

$$\begin{bmatrix} \dfrac{\partial P_{\text{Slack}}}{\partial P_G} & \dfrac{\partial P_{\text{Slack}}}{\partial Q_G} \end{bmatrix} \begin{bmatrix} \dfrac{\partial P}{\partial \theta} & \dfrac{\partial P}{\partial V} \\ \dfrac{\partial Q}{\partial \theta} & \dfrac{\partial P}{\partial V} \end{bmatrix} = \begin{bmatrix} \dfrac{\partial P_{\text{Slack}}}{\partial \theta} & \dfrac{\partial P_{\text{Slack}}}{\partial V} \end{bmatrix} \tag{5.39}$$

or

$$\begin{bmatrix} \dfrac{\partial P}{\partial \theta} & \dfrac{\partial P}{\partial V} \\ \dfrac{\partial Q}{\partial \theta} & \dfrac{\partial P}{\partial V} \end{bmatrix}^{t} \begin{bmatrix} \dfrac{\partial P_{\text{Slack}}}{\partial P_G} \\ \dfrac{\partial P_{\text{Slack}}}{\partial Q_G} \end{bmatrix} = \begin{bmatrix} \dfrac{\partial P_{\text{Slack}}}{\partial \theta} \\ \dfrac{\partial P_{\text{Slack}}}{\partial V} \end{bmatrix} \tag{5.40}$$

A DC power flow can be used to estimate the penalization factor at the reference node. Neglecting the shunts, the branch losses are

$$\mathbf{S}_{ij} = \left(\mathbf{V}_i - \mathbf{V}_j\right)\mathbf{I}_{ij}^* = \left(\mathbf{V}_i - \mathbf{V}_j\right)\left[\left(\mathbf{V}_i - \mathbf{V}_j\right)\mathbf{Y}_{ij}\right]^*$$

$$\mathbf{S}_{ij} = \left(\mathbf{V}_i - \mathbf{V}_j\right)\left(\mathbf{V}_i^* - \mathbf{V}_j^*\right)\mathbf{Y}_{ij}^*$$

$$\mathbf{S}_{ij} = \left[V_i^2 + V_j^2 - 2V_iV_j\cos(\theta_i - \theta_j)\right]\mathbf{Y}_{ij}^* \tag{5.41}$$

Then, the active losses through branch ij are

$$P_{ij} = Re(\mathbf{S}_{ij}) = \left[V_i^2 + V_j^2 - 2V_iV_j\cos(\theta_i - \theta_j)\right]G_{ij} \tag{5.42}$$

And, the total incremental losses can be estimated as

$$\Delta P_{\mathrm{L}} = \sum_{\forall\text{lines}} \frac{\partial P_{\text{lines}}}{\partial\theta_j} \frac{\partial\theta_j}{\partial P_i}\Delta P_i \tag{5.43}$$

resulting in

$$\frac{\Delta P_{\mathrm{L}}}{\Delta P_i} = \sum_{\forall\text{bus}\,j}\left[\sum_{\forall\text{lines}\Rightarrow j}\frac{\partial P_{\text{lines}}}{\partial\theta_j}\right]\frac{\partial\theta_j}{\partial P_i} \tag{5.44}$$

$$\overline{[B']}^t\left[\frac{\partial P_{\mathrm{L}}}{\partial P}\right] = \left[\frac{\partial P_{\mathrm{L}}}{\partial\theta}\right] \tag{5.45}$$

B matrix is the system susceptance matrix:

$$[B] = Im\,[Y] \tag{5.46}$$

Matrix $\mathbf{B}'$ is calculated considering the network without the column and row related to the reference node, considering that all voltage are 1.0 p.u., the losses are

$$\frac{\partial P_{\mathrm{L}}}{\partial\theta_j} = \sum_{\forall\text{lines}\Rightarrow j}\frac{\partial P_{\text{branch}}}{\partial\theta_j} = \sum_{\forall\text{branch}\Rightarrow j} -2\sin(\theta_i - \theta_j)G_{ij} \tag{5.47}$$

5.8 NEW ADVANCES IN ECONOMIC DISPATCH CALCULATIONS

Nowadays, there are several new challenges related to the ED calculations. First, the system sizes are increasing which poses a significant challenge for the existing algorithms, particularly for real-time applications. Furthermore, the inclusion of renewable resources brings uncertainties to the current formulations. Besides, the demand for emission controls requires the addition of new constraints into the ED formulation. One way to solve this kind of problem is to formulate the problem as an optimal power flow (OPF), which will be discussed next. Moreover, many researchers proposed the application of stochastic search algorithms such as genetic algorithms (GAs), evolutionary programming (EP), and simulated annealing (SA). These methods seem to be very useful in solving nonlinear ED problems without any restrictions on the shape of the cost curves. They provide a fast and reasonable solution (suboptimal or near globally optimal). Both GA and EP can provide a near global solution. One drawback of these methods is related to the performance, which poses challenges for these methods to become a good alternative for real-time applications. Hybrid methods that combine two or more of these methods are also proposed. In addition, methods based on particle swarm optimization (PSO) is suggested to solve complex ED problems. PSO is gaining a lot of attention in various power system applications. Among the non-linear programming methods, the sequential quadratic programming (SQP) methodology is considered to outperform other nonlinear programming methods in the areas of efficiency, accuracy, and percentage of successful solutions, for ED type of applications for a large-scale system.

CHAPTER END PROBLEMS

5.1 For a two-generator system with the following fuel cost curves:

$$C(P_{G_1}) = 80 + 8P_{G_1} + 0.024P_{G_1}^2$$

$$C(P_{G_2}) = 120 + 6P_{G_2} + 0.04P_{G_2}^2$$

where $C(P_G)$[MBtu/h], P_G[MW], $P_{G1}^{Min} = P_{G2}^{Min} = 20$ MW, $P_{G1}^{Max} = P_{G2}^{Max} = 120$ MW.
 Plot hourly consumption and the incremental cost curve as a function of the unit power. Consider that the cost is $1.5/MBtu. Calculate the economic dispatch considering the following loads: 50 MW, 100 MW, 150 MW, and 120 MW. Use the different methods described in the chapter.

5.2 Calculate the performance of a generator whose specific heat rate is given by Table 5.2.

5.3 Calculate the lossless economic dispatch considering the generator data described in Table 5.3 and a 500 MW load using the different methodologies explained in the Chapter.

5.4 Consider the system described by Figure 5.14, line parameters, generation, and loads given in Tables 5.4 and 5.5, and calculate the penalization value at each generation nodes.

Table 5.2 Specific Consumption

P_G (MW)	C (kcal/kWh)
5	6038.5
10	4005.0
15	3341.5
20	3020.5
25	2836.5
30	2721.0
40	2592.8
50	2533.0
60	2507.5
70	2501.6
75	2503.5

Table 5.3 Generator Data

Unit ID	a [kcal/h]	b [kcal/MWh]	c [kcal/MW^{2}h]	P^{Min} [MW]	P^{Max} [MW]
1	17787	2227	9.60	150	300
2	13387	2373	14.20	150	250
3	14931	3000	20	100	200

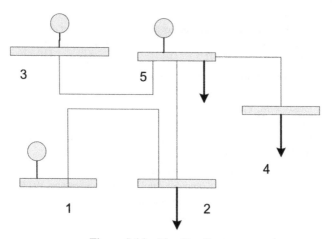

Figure 5.14 Five-Bus System.

Table 5.4 Line Parameters

From Node	To Node	R	X
1	2	0.01	0.03
2	5	0.09	0.25
5	3	0.03	0.06
5	4	0.03	0.05

Table 5.5 Generation and Load parameters

Node	Generation (MW)	load (MW)
1	100	—
2	—	120
3	150	—
4	—	200

FURTHER READING

1. Davison GR. Dividing load between units. *Electrical World* 1922;80(26):1385–1387.

2. Wilstam A. Dividing load economically among power plants by use of the kilowatt–Killowatt-hour curve. *Journal of the AIEE* 1928;47(6):430–432.

3. Estrada H. Economical load allocations. *Electrical World* October 11, 1930;96:685–690.

4. Stahl CE. Load division in interconnections. *Electrical World* March 1, 1930;95:434–438.

5. Hahn G. Load division by the increment method. *Power* June, 1931.

6. Kirchmayer LK. *Economic Operation of Power Systems*, 1st edition. New York: John Wiley & Sons; 1958.

7. El-hawary ME, Mansour SY. Performance evaluation of parameter estimation algorithms for economic operation of power systems. *IEEE Transactions on Power Apparatus and Systems* 1982;PAS-101(3):574–582.

8. Liang Z, Glover JD. Improved cost functions for economic dispatch computations. *IEEE Transactions on Power Systems* 1991;6(2):821–829.

9. Noyola AH, Grady WM, Viviani GL. An optimized procedure for determining incremental heat rate characteristics. *IEEE Transactions on Power Systems* 1990;5(2):376–383.

10. Wolfe PM. Methods of nonlinear programming. In: *Recent Advances in Mathematical Programming*, Graves RL, Wolfe PM, editors. McGraw-Hill; 1963.

11. Podmore, R. A simplified and improved method for calculating transmission loss formulas. In: Proceedings of Power Industry Computer Applications Conference, 1973, pp. 428–432.

12. Shoults RR, Grady WM, Helmick S. An efficient method for computing loss formula coefficients based upon the method of least squares. *IEEE Transactions on Power Apparatus and Systems* 1979;PAS-98(6):2144–2152.

13. Chang YC, Yang WT, Liu CC. A new method for calculating loss coefficients. *IEEE Transactions on Power Systems* 1994;9(3):1665–1671.

Chapter 6

Optimal Power Flow

In the previous chapter, the economic dispatch problem concerns only with the optimal schedule of real power by minimizing the production cost including the cost of the network security. In a more general optimal power flow problem, both real and reactive power schedules can be optimized. In fact, other system parameters such as generator voltages, transformer taps, and reactive power injection can also be optimized. This general problem can be expanded to other types of problems.

6.1 INTRODUCTION

Since the early 1960s, the transmission network part of electric power systems all over the world began to operate at or nearly at their maximum capacities. This problem necessitated the consideration and the inclusion of security constraints into economic dispatch problems which later became one of the optimal power flow (OPF) problems.

Initially, the OPF was conceived as an extension of the conventional ED, where the ED and power flow problems are solved simultaneously. This early model of OPF can be formulated as a nonlinear problem whose objective function is to determine the control variables and state variables that minimize the generation costs subject to the power balance equation and the transmission network constraints to ensure the twin goals of economic and secure system operation.

As the first step in formulating an OPF problem, the following must be determined first: variables, constraints, and objective function.

- **Variables**

  a. Control variables (u): Active and reactive power, generator voltage magnitudes, transformer tap, phase shifters

Electricity Markets: Theories and Applications, First Edition. Jeremy Lin and Fernando H. Magnago.
© 2017 by The Institute of Electrical and Electronics Engineers, Inc. Published 2017 by John Wiley & Sons, Inc.

b. State variables (x): Bus voltage magnitudes, relative bus voltage angle difference with respect to the reference angle

- **Constraints**

Both equality and inequality constraints are present in this problem. Equality constraints are related to the power flow balance; inequality constraints are related to the upper and lower limits of systems equipments such as transmission lines, generators or transformers.

- **Objective Function**

A generic OPF problem can have several objective functions such as fuel cost minimization, active or reactive loss minimization, and minimum control movement. This flexibility makes it possible to use the OPF for different types of problems. These objective functions are further elaborated below.

1. **Economic Dispatch**: To minimize the operation cost including the cost of the network security (network facilities operating within their limits).
2. **Preventive Dispatch**: To include contingency constraints into the problem and adjust the controls in order to keep the voltages and branch flows within the limits even with the presence of a contingency.
3. **Corrective Dispatch**: During an emergency, if a system equipment is over its limit, or if a voltage violation occurs, the OPF solution can suggest the system operator the necessary system adjustments to ensure that all system elements are operated within their steady-state normal limits.
4. **Volt–Var Optimization**: An OPF can be used periodically to get the optimal operating point of generator voltages, transformer taps, and injection of reactive power.
5. **Spot Price Calculation**: An OPF can be used to calculate the energy price per node in the system on a fixed interval basis under the electricity market regime.

6.2 GENERIC FORMULATION FOR OPF PROBLEM

The canonical form is

$$
\begin{aligned}
\text{minimize} \quad & f(\mathbf{x}, \mathbf{u}) \\
\text{subject to} \quad & \\
& \mathbf{g}(\mathbf{x}, \mathbf{u}, \mathbf{p}) = \mathbf{0} \Leftrightarrow \lambda \\
& \mathbf{h}(\mathbf{x}, \mathbf{u}, \mathbf{p}) \leq \mathbf{0} \Longleftrightarrow \mu
\end{aligned}
\tag{6.1}
$$

where

$\mathbf{x}$ = state variable vector
$\mathbf{u}$ = control variable vector
$\mathbf{p}$ = parameter vector
f = objective function
$\mathbf{g}$ = power flow equations
$\mathbf{h}$ = operation and physical constraints
λ = Lagrange multiplier for the power balance
μ = Lagrange multiplier for operational and physical constraints

The Lagrange function can be defined as

$$\pounds = f(\mathbf{x}, \mathbf{u}, \mathbf{p}) + \lambda^t \mathbf{g}(\mathbf{x}, \mathbf{u}, \mathbf{p}) + \mu^t \mathbf{h}(\mathbf{x}, \mathbf{u}, \mathbf{p})$$

The first-order optimality conditions (KKT) requires

$$\frac{\partial \mathbf{f}}{\partial \mathbf{x}} + \left[\frac{\partial \mathbf{g}}{\partial \mathbf{x}}\right]^t \lambda + \left[\frac{\partial \mathbf{h}}{\partial \mathbf{x}}\right]^t \mu = 0$$

$$\frac{\partial \mathbf{f}}{\partial \mathbf{u}} + \left[\frac{\partial \mathbf{g}}{\partial \mathbf{u}}\right]^t \lambda + \left[\frac{\partial \mathbf{h}}{\partial \mathbf{u}}\right]^t \mu = 0$$

$$\mathbf{g}(\mathbf{x}, \mathbf{u}, \mathbf{p}) = 0 \qquad (6.2)$$

$$\mathbf{h}(\mathbf{x}, \mathbf{u}, \mathbf{p}) \leq 0$$

$$\mu_i \quad h_i(x, u, p) = 0$$

$$\mu_i \geq 0$$

The first equation in Eq. (6.2) is the Lagrangian gradient with respect to the state variables (x). And the second equation is the Lagrangian gradient with respect to the control variables (u). These equations are equal to zero at the optimal solution point. Third and fourth equations are constraints to the problem. The last set of equations are known as complementary equations which provide a way to handle the restrictions as active or nonactive.

6.3 SOLUTION METHODS FOR OPF PROBLEM

Since 1930s, when the first techniques for minimum fuel cost calculation were developed, several methods have been developed to obtain an optimal operation of the power system. The equal incremental cost method known as EICC is considered the predecessor of the OPF. The first OPF methodologies, developed in 1960s, used an iterative process to include nonlinear constraints in the problem formulation. Since then, there have been developments of many techniques which can be classified into the following groups:

1. Heuristic methods
2. Primal methods
3. Penalty and barrier methods
4. Linear programming-based methods

Among all these methods, the methods that are most-widely used are described next.

6.3.1 Primal Methods

The primal methods attempt to find the solution in the feasible region. The need to obtain an initial feasible point is the main drawback of the method. For problems with equality constraints, it is necessary to preserve the feasible region during the iterative process. In 1968, Dommel and Tinney proposed a gradient-based method which starts as an initial solution, a feasible power flow without the inequality constraints. Then, these constraints are later added to the OPF problem using penalty functions. Mathematically, OPF problem based on gradient method can be summarized as

$$\text{minimize} \quad f(\mathbf{x}, \mathbf{u}) \tag{6.3}$$

subject to

$$g(\mathbf{x}, \mathbf{u}, \mathbf{p}) = \mathbf{0} \iff \lambda \tag{6.4}$$

The Lagrangian is

$$\pounds = f(\mathbf{x}, \mathbf{u}, \mathbf{p}) + \lambda^t \mathbf{g}(\mathbf{x}, \mathbf{u}, \mathbf{p}) \tag{6.5}$$

and the KKT conditions are

$$\nabla \pounds_x = \frac{\partial \mathbf{f}}{\partial \mathbf{x}} + \left[\frac{\partial \mathbf{g}}{\partial \mathbf{x}}\right]^t \lambda = 0 \tag{6.6}$$

$$\nabla \pounds_u = \frac{\partial \mathbf{f}}{\partial \mathbf{u}} + \left[\frac{\partial \mathbf{g}}{\partial \mathbf{u}}\right]^t \lambda = 0$$

$$\mathbf{g}(\mathbf{x}, \mathbf{u}, \mathbf{p}) = \mathbf{0}$$

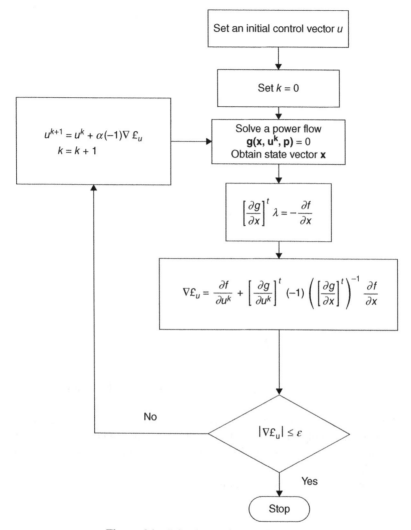

Figure 6.1   Primal-Based OPF Algorithm.

The Lagrangian gradient with respect to the control variables ($\nabla \pounds_u$) gives the direction of the maximum variation, and must be zero at the optimal point. Figure 6.1 illustrates a simple primal-based OPF algorithm that describes this technique. The critical step in this algorithm is the control vector update because it depends on the selection of the empirical value α. A small value of α may guarantee the convergence but it may be slow to achieve that convergence. A large value of α can speed up the solution convergence, but it may oscillate. In addition to the use of penalty

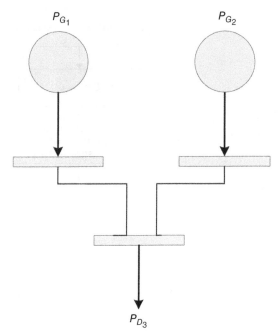

Figure 6.2   Three Bus System Example.

functions, this estimation of the step size α makes this method not suitable for real-time applications.

To illustrate this method, let us consider the small system described in Figure 6.2 with the following data: $P_{G_2} = 1.7$ p.u., $P_{L_3} = 2$ p.u., and $Q_{G_3} = 1.0$ p.u., line parameters are $y_{13} = 4 - j10$ p.u. and $y_{23} = 4 - j5$ p.u. Assume the active loss minimization as the problem objective.

The control variables, states, and parameters are

$$\mathbf{u} = \begin{bmatrix} |E_1| \\ |E_2| \end{bmatrix}; \quad \mathbf{x} = \begin{bmatrix} \theta_2 \\ \theta_3 \\ |E_3| \end{bmatrix}; \quad \mathbf{p} = \begin{bmatrix} \theta_1 \\ P_2 \\ P_3 \\ Q_3 \end{bmatrix} \tag{6.7}$$

The objective function is

$$f(\mathbf{x}, \mathbf{u}) = P_1(\theta_2, \theta_3, |E_3|; |E_1|, |E_2|) \tag{6.8}$$

Since active power P_2 and P_3 are fixed, the loss minimization is performed by changing the slack generation:

$$\frac{\partial \mathbf{f}}{\partial \mathbf{x}} = \begin{bmatrix} \dfrac{\partial P_1}{\partial \theta_2} \\[2mm] \dfrac{\partial P_1}{\partial \theta_3} \\[2mm] \dfrac{\partial P_1}{\partial |E_3|} \end{bmatrix}; \quad \frac{\partial \mathbf{g}}{\partial \mathbf{x}} = \begin{bmatrix} \dfrac{\partial P_2}{\partial \theta_2} & \dfrac{\partial P_2}{\partial \theta_3} & \dfrac{\partial P_2}{\partial |E_3|} \\[2mm] \dfrac{\partial P_3}{\partial \theta_2} & \dfrac{\partial P_3}{\partial \theta_3} & \dfrac{\partial P_3}{\partial |E_3|} \\[2mm] \dfrac{\partial Q_3}{\partial \theta_2} & \dfrac{\partial Q_3}{\partial \theta_3} & \dfrac{\partial Q_3}{\partial |E_3|} \end{bmatrix}; \quad \frac{\partial \mathbf{f}}{\partial \mathbf{u}} = \begin{bmatrix} \dfrac{\partial P_1}{\partial |E_1|} \\[2mm] \dfrac{\partial P_1}{\partial |E_2|} \end{bmatrix}$$

$$\frac{\partial \mathbf{g}}{\partial \mathbf{u}} = \begin{bmatrix} \dfrac{\partial P_2}{\partial |E_1|} & \dfrac{\partial P_2}{\partial |E_2|} \\[2mm] \dfrac{\partial P_3}{\partial |E_1|} & \dfrac{\partial P_3}{\partial |E_2|} \\[2mm] \dfrac{\partial Q_3}{\partial |E_1|} & \dfrac{\partial Q_3}{\partial |E_2|} \end{bmatrix}; \quad \mathbf{g(x, u)} = \begin{bmatrix} P_2(|E|, \theta) - P_2 \\[1mm] P_3(|E|, \theta) - P_3 \\[1mm] Q_3(|E|, \theta) - Q_3 \end{bmatrix}$$

First, the initial values of the control vector are set as an initial calculation:

$$\mathbf{u}^0 = \begin{bmatrix} |E_1|^0 \\[1mm] |E_2|^0 \end{bmatrix} = \begin{bmatrix} 1.1 \\[1mm] 0.9 \end{bmatrix}$$

Results from the first power flow are total active power losses $= 0.3906$ p.u., slack active power $= 0.6906$ p.u. Then,

$$\frac{\partial \mathbf{f}}{\partial \mathbf{x}} = \begin{bmatrix} 0.0 \\ 4.36 \\ 4.14 \end{bmatrix}; \quad \frac{\partial \mathbf{g}}{\partial \mathbf{x}} = \begin{bmatrix} 8.14 & 8.14 & 1.54 \\ 6.96 & 12.0 & 3.85 \\ -4.5 & -7.85 & 10.0 \end{bmatrix}$$

and the system results become

$$\begin{bmatrix} 8.14 & 8.14 & 1.54 \\ 6.96 & 12.0 & 3.85 \\ -4.5 & -7.85 & 10.0 \end{bmatrix}^t \lambda = - \begin{bmatrix} 0.0 \\ 4.36 \\ 4.14 \end{bmatrix}$$

λ values are

$$\begin{bmatrix} \lambda_1 \\ \lambda_2 \\ \lambda_3 \end{bmatrix} = \begin{bmatrix} 0.743 \\ -0.98 \\ -0.154 \end{bmatrix}$$

In order to calculate the Lagrangian variations, we have

$$\frac{\partial f}{\partial u} = \begin{bmatrix} 5.533 \\ 0.0 \end{bmatrix} ; \quad \frac{\partial g^t}{\partial u} = \begin{bmatrix} 0.0 & 3.354 & 5.0 \\ 4.94 & 4.5 & 6.96 \end{bmatrix}$$

which gives the following results:

$$\nabla \pounds_u = \begin{bmatrix} 5.533 \\ 0.0 \end{bmatrix} + \begin{bmatrix} 0.0 & 3.354 & 5.0 \\ 4.94 & 4.5 & 6.96 \end{bmatrix} \begin{bmatrix} 0.743 \\ -0.98 \\ -0.154 \end{bmatrix} = \begin{bmatrix} 2.25 \\ -1.78 \end{bmatrix}$$

Setting $\alpha = 0.03$, the control vector update is

$$u^1 = \begin{bmatrix} |E_1|^1 \\ |E_2|^1 \end{bmatrix} = \begin{bmatrix} 1.1 \\ 0.9 \end{bmatrix} - 0.03 \begin{bmatrix} 2.25 \\ -1.78 \end{bmatrix} = \begin{bmatrix} 0.95 \\ 1.03 \end{bmatrix}$$

Results from the second iteration are total active power losses $= 0.2380$ p.u., slack active power $= 0.5380$ p.u., and

$$u^2 = \begin{bmatrix} |E_1|^2 \\ |E_2|^2 \end{bmatrix} = \begin{bmatrix} 0.86 \\ 0.86 \end{bmatrix}$$

This iterative process continues until there is no more loss variations. Nevertheless, the losses may increase, rather than decrease, during the process, then α value may need to be tuned, which represents a disadvantage of this method, particularly for real-time applications.

6.3.2 Penalty and Barrier Methods

In 1984, Sun et al. presented an algorithm based on the Newton–Raphson method by exploiting the sparsity of the *Hessian* matrix resulting from the Lagrangian quadratic approximation. The main challenge of this algorithm was to identify the active constraints efficiently.

$$\text{Minimize} \quad f(\mathbf{x}, \mathbf{u}) \tag{6.9}$$

subject to

$$\mathbf{g}(\mathbf{x}, \mathbf{u}, \mathbf{p}) = \mathbf{0} \Leftrightarrow \lambda$$
$$\mathbf{h}(\mathbf{x}, \mathbf{u}, \mathbf{p}) \leq \mathbf{0} \Longleftrightarrow \mu$$

The Lagrangian function is

$$\pounds = f(\mathbf{x}, \mathbf{u}, \mathbf{p}) + \lambda^t \mathbf{g}(\mathbf{x}, \mathbf{u}, \mathbf{p}) + \mu^t \mathbf{h}(\mathbf{x}, \mathbf{u}, \mathbf{p}), \tag{6.10}$$

where $z = [\mathbf{x} \ \mathbf{u} \ \mathbf{p} \ \lambda \ \mu]^t$ include the state variables, control variables, parameters, and Lagrange multipliers for the balance equation and the restrictions. The restrictions include only the active power restrictions. The Lagrange gradient and the Hessian are

$$\nabla \pounds(z) = \left[\frac{\partial \pounds(z)}{\partial z_i} \right] \tag{6.11}$$

$$\mathbf{H} = \left[\frac{\partial^2 \pounds(z)}{\partial z_i \partial z_j} \right] = \begin{bmatrix} \dfrac{\partial^2 \pounds(z)}{\partial x_i \partial x_j} & \dfrac{\partial^2 \pounds(z)}{\partial x_i \partial u_j} & \dfrac{\partial^2 \pounds(z)}{\partial x_i \partial \lambda_j} & \dfrac{\partial^2 \pounds(z)}{\partial x_i \partial \mu_j} \\[2mm] \dfrac{\partial^2 \pounds(z)}{\partial u_i \partial x_j} & \dfrac{\partial^2 \pounds(z)}{\partial u_i \partial u_j} & \dfrac{\partial^2 \pounds(z)}{\partial u_i \partial \lambda_j} & \dfrac{\partial^2 \pounds(z)}{\partial u_i \partial \mu_j} \\[2mm] \dfrac{\partial^2 \pounds(z)}{\partial \lambda_i \partial x_j} & \dfrac{\partial^2 \pounds(z)}{\partial \lambda_i \partial u_j} & 0 & 0 \\[2mm] \dfrac{\partial^2 \pounds(z)}{\partial \mu_i \partial x_j} & \dfrac{\partial^2 \pounds(z)}{\partial \mu_i \partial u_j} & 0 & 0 \end{bmatrix} \tag{6.12}$$

The OPF can be solved by setting $\nabla \pounds(z) = 0$ using the Newton–Raphson method. The algorithmic steps are shown in Figure 6.3.

The Lagrangian includes only the active power inequality constraints. Consequently, their correct selection is critical. The sensitivity theorem establishes that the multipliers oppose the derivative of the objective function with respect to the active constraints limits. Therefore, if the multiplier is positive, the objective function decreases, and the constraint must remain active. This criterion is then used to select the active constraints within the programming cycle. If there is no feasible solution, it is necessary to relax the constraints by adding penalty functions. Ideally, they are small and increase rapidly if a limit is violated. For the Newton method, it is

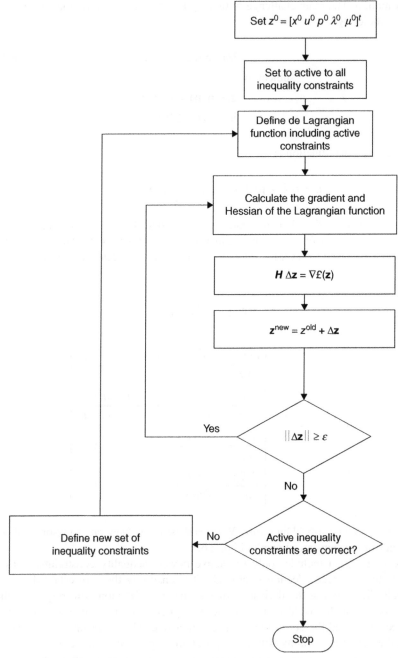

Figure 6.3 Penalty and Barrier Method.

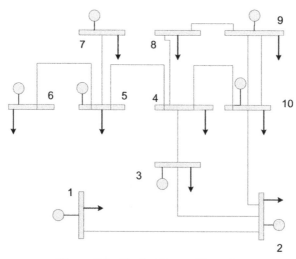

Figure 6.4 Ten Bus System Example.

convenient to use quadratic penalty functions. For example, if the relaxed constraints are the bus voltage constraints represented as follow:

$$V_i^{\text{Min}} \leq V_i \leq V_i^{\text{Max}}$$

The following penalty functions are defined for each bus voltage module:

$$W_i = \begin{cases} k\left(V_i^{\text{Min}} - V_i\right)^2; & \text{if} \quad V_i < V_i^{\text{Min}} \\ 0; & \text{if} \quad V_i^{\text{Min}} \leq V_i \leq V_i^{\text{Max}} \\ k\left(V_i - V_i^{\text{Max}}\right)^2; & \text{if} \quad V_i > V_i^{\text{Max}} \end{cases}$$

The penalty function is null if the voltage limits are not violated. Penalty function starts to increase if they are violated. Adjusting parameter k can be selected for the priority of this constraint.

To illustrate the method, let us consider the one-line diagram of a simplified transmission system represented in Figure 6.4. It is a 500 kV system with a total capacity of 16,750 MW. The parameters are described in Tables 6.1 and 6.2.

All voltage limits are set to ±5% of the nominal values. The objective function is formulated as the minimization of the total generation cost:

$$\sum_{i=1}^{8} C(P_i) = \sum_{i=1}^{8} a_i + b_i P_i + c_i P_i^2$$

The results of the OPF are total cost = 91,743.83 $/h, active losses = 114.52 MW, and reactive losses = 1542.74 MVAr. Tables 6.3–6.8 show additional results.

Table 6.1 Load

Bus	P (MW)	Q (MVAr)
1	300	100
2	5000	2000
3	50	10
4	700	140
5	500	140
6	200	60
7	300	90
8	300	110
9	1000	300
10	200	60

Table 6.2 Cost Functions

Generator	a [kcal/h]	b [kcal/MWh]	c [kcal/MW^{2}h]	P^{Min} [MW]	P^{Max} [MW]
1	0.0	1.0	0.003	0.0	5800
2	0.0	9.5	0.002	0.0	5000
3	0.0	1.0	0.050	300	350
5	0.0	10.0	0.050	600	650
6	0.0	9.0	0.010	0.0	1300
7	0.0	7.1	0.040	0.0	1000
9	0.0	1.0	0.001	0.0	1700
10	0.0	1.0	0.001	900	950

6.3.3 Linear Programming-Based Methods

Another alternative to solve an OPF problem is to linearize the objective function as well as the equality and inequality constraints around the operating point and solve

Table 6.3 Optimal Dispatch

Bus	Voltage (p.u.)	Angle	MW	MVAr
1	1.050	0.0	2521.42	559.78
2	1.050	−32.51	2055.14	2878.70
3	1.050	−29.53	300.0	89.88
4	1.036	−28.54	—	—
5	1.050	−26.35	600.0	165.05
6	1.050	−21.49	405.40	29.51
7	1.044	−30.34	132.56	85.30
8	1.024	−23.47	—	—
9	1.048	−17.63	1700.0	371.42
10	1.050	−22.66	950.0	127.24

Table 6.4   Active Power Flows

From	To	From P (MW)	To P (MW)	Losses P (MW)
1	2	2221.42	−2127.83	93.59
2	3	−360.96	362.15	1.19
3	4	−112.15	112.36	0.21
4	5	−134.21	134.75	0.53
5	6	−203.48	205.40	1.92
5	7	168.74	−167.44	1.29
4	8	−184.68	186.04	1.36
10	2	463.87	−456.07	7.8
10	9	−209.61	211.67	2.06
4	10	−493.47	495.74	2.27
8	9	−486.04	488.33	2.29

Table 6.5   Reactive Power Flows

From	To	From Q (MVAr)	To Q (MVAr)	Losses Q (MVAr)
1	2	459.78	780.76	1268.10
2	3	26.78	−18.98	18.82
3	4	98.78	−106.38	3.36
4	5	−42.94	26.81	5.61
5	6	14.78	−30.49	17.36
5	7	−16.54	−4.70	11.66
4	8	33.93	−43.59	16.85
10	2	−19.47	71.16	79.25
10	9	21.58	−35.99	18.59
4	10	−24.61	65.13	51.40
8	9	−66.41	107.41	51.73

Table 6.6   Voltage Constraints

Bus	V^{Min}	μ_V^{Min}	V	V^{Max}	μ_V^{Max}
1	0.950	—	1.050	1.050	2241.6
2	0.950	—	1.050	1.050	1288.2
3	0.950	—	1.050	1.050	205.8
5	0.950	—	1.050	1.050	67
6	0.950	—	1.050	1.050	114.5
10	0.950	—	1.050	1.050	402

Table 6.7 Generation Constraints

Bus	P^{Min}	μ_P^{Min}	P	P^{Max}	μ_P^{Max}
3	300.0	13.4	300.0	350.0	—
5	600.0	52.6	600.0	650.0	—
9	0.0	—	1700	1.700	12.2
10	900.0	—	950.0	950.0	14.2

Table 6.8 Flow Constraints

From	S_{4-10}	$\mu_{S_{4-10}}$	S^{Max}	S_{10-4}	$\mu_{S_{10-4}}$	To
4	494.08	—	500.0	500.0	0.159	10
8	490.55	—	500.0	500.0	0.940	9

it by using a linear programming method. This formulation is adequate as soon as the linearization can be done without sacrificing precision. Then, the only remaining variable is the control vector which can be formulated similarly to the expression described by Eq. (6.13). The basic OPF algorithm based on LP method is shown in Figure 6.5.

To illustrate the method, consider a convex and quadratic objective function which represents the production cost. Then, it can be approximated by a piecewise linear function:

$$C_i(P_i) = C_i\left(P_i^{\text{Min}}\right) + \sum_{k=1}^{N} s_{ik} P_{ik}$$

$$P_{ik}^- \le P_{ik} \le P_{ik}^+; \quad k = 1, \ldots, N$$

$$P_i = P_i^{\text{Min}} + \sum_{k=1}^{N} P_{ik}$$

$$P_i^{\text{Min}} \le P_i \le P_i^{\text{Max}}$$

The control variables in this case are P_{ik}. To linearize the problem, consider the balance equations:

$$P_{\text{Gen}} = P_{\text{Demand}} + P_{\text{Loss}}$$

$$Q_{\text{Gen}} = Q_{\text{Demand}} + Q_{\text{Loss}}$$

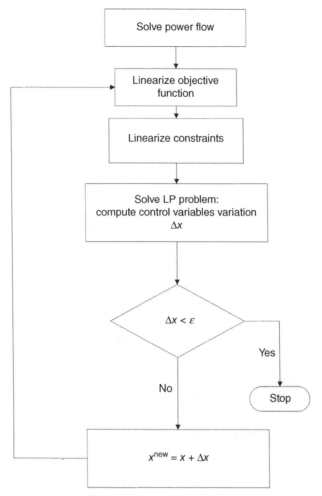

Figure 6.5   LP Based Method.

The derivatives with respect to the control variables are

$$\sum_u \left(\frac{\partial P_{\text{Gen}}}{\partial u} \right) \Delta u - \sum_u \left(\frac{\partial P_{\text{Demand}}}{\partial u} \right) \Delta u - \sum_u \left(\frac{\partial P_{\text{Loss}}}{\partial u} \right) \Delta u = 0$$

$$\sum_u \left(\frac{\partial Q_{\text{Gen}}}{\partial u} \right) \Delta u - \sum_u \left(\frac{\partial Q_{\text{Demand}}}{\partial u} \right) \Delta u - \sum_u \left(\frac{\partial Q_{\text{Loss}}}{\partial u} \right) \Delta u = 0$$

replacing $\Delta u = u - u^0$:

$$\sum_u \left(\frac{\partial P_{\text{Gen}}}{\partial u}\right) u - \sum_u \left(\frac{\partial P_{\text{Demand}}}{\partial u}\right) u - \sum_u \left(\frac{\partial P_{\text{Loss}}}{\partial u}\right) u = K_{\text{P}}$$

$$\sum_u \left(\frac{\partial Q_{\text{Gen}}}{\partial u}\right) u - \sum_u \left(\frac{\partial Q_{\text{Demand}}}{\partial u}\right) u - \sum_u \left(\frac{\partial Q_{\text{Loss}}}{\partial u}\right) u = K_{\text{Q}}$$

The left side is constant since u^0 represents the control value at the operation point. The inequality constraints must be linearized as well, for the transmission line limits:

$$P_{ij} \leq P_{ij}^{\text{Max}}$$

Expanding the power flow using *Taylor series*:

$$P_{ij} \cong P_{ij}^0 + \sum_u \left(\frac{\partial P_{ij}}{\partial u}\right) \Delta u$$

and replacing $\Delta u = u - u^0$:

$$\sum_u \left(\frac{\partial P_{ij}}{\partial u}\right) u \leq P_{ij}^{\text{Max}} - P_{ij}^0 - \sum_u \left(\frac{\partial P_{ij}}{\partial u}\right) u^0$$

Similar analysis can be done for all constraints. For this type of methods, it is important to be familiar with solving a problem based on the Linear Programming formulation. Therefore, LP method is described next.

6.3.4 Linear Programming

In general, one of the problems in the optimization problem formulation is related to the representation of the inequality constraints. One alternative to solve this problem is to linearize the constraints and the objective function near the operating point and use linear programming techniques to achieve the optimal solution.

6.3.4.1 Definition The general form of a linear programming formulation can be written as

$$\text{minimize} \quad f(x) = c^T x \tag{6.13}$$

$$\text{subject to}$$

$$Ax = b$$

$$x \geq 0,$$

where c is an n-vector, A is a $(m \times n)$ matrix, $(m < n)$, and b is an m-vector.

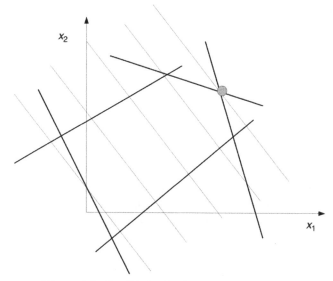

Figure 6.6   Feasible Region in Linear Programming.

6.3.4.2 Feasible Solution   Any solution that satisfies all the constraints is known as the *feasible solution*. The set of all feasible solutions is called the *feasible region*. If the feasible region is bounded, there always exists an *optimal solution*. Geometrically, the feasible region is a *convex polyhedron* and the minimum solution must be placed at one of its vertices.

6.3.4.3 Optimal Solution   From all the feasible solutions, the solution that satisfies the inequality $\{f(x^*) > f(x)\}$, is known as the optimal solution. This optimal solution is always a *global optimum*. If there are more than one optimal solution, these solutions are a linear combination of each other. The optimal solution is always at an extreme point of the feasible region. Figure 6.6 shows an example of a feasible region and the optimal point.

6.3.4.4 Numerical Methods for LP Problems   The primary method that can be used to find the optimal solution for LP problems is to move from one extreme to another extreme based on certain rules until the lowest objective value is found. The standard method used to solve a linear programming problem is known as the *simplex method*. To explain this method, the linear model can be reformulated by dividing the solution vector into the basic solution and the non-basic solution:

$$[B \quad N] \begin{bmatrix} x_B \\ 0 \end{bmatrix} = b, \tag{6.14}$$

where the basic variable is

$$x_b = B^{-1}b \tag{6.15}$$

the objective function can be represented as follows:

$$c^T x = [c_B \quad c_N] \begin{bmatrix} x_B \\ 0 \end{bmatrix} = c_B x_B = c_B B^{-1} b \tag{6.16}$$

Based on the value of the cost vector, it is possible to take decisions regarding the objective function and the non-basic variables, and move to another feasible solution and check if the cost is lowered due to this modification. Let us assume that the solution moved from $x = (x_B 0)$ to another solution $x^k = (x_B x_N)$. The new basic variable is

$$x_b^k = x_B - B^{-1} N x_N \tag{6.17}$$

and the new objective function is

$$[c_B \quad c_N] \begin{bmatrix} x_B - B^{-1} N x_n \\ x_N \end{bmatrix} = \left(c_N - c_B B^{-1} N \right) x_N + c_B x_B \tag{6.18}$$

From the inspection of the equation, it is clear that the objective function can be modified by changing the term $(c_N - c_B B^{-1} N)$. This term is known as the *residual*. If all the residual elements are non-zero, then the optimal value is reached.

Based on these equations, the following statements are the basic steps of the Simplex algorithm.

1. Find an initial vertex from the feasible set, which is called basic feasible solution.

2. Select a non-basic variable that gives the fastest rate of increase in the objective function value. Such a non-basic variable can be the most negative component of the residual vector.

3. Choose the variable leaving the basic set. To perform this task, first select the column z corresponding to the entering variable in the non basic set N, then, solve the system $By = z$. Calculate the ratio values from $\frac{x_b}{y}$, and select the variable from vector x_B which gives the smallest ratio.

4. Update the equations and check the residual to see if it is optimal; otherwise continue with the process until optimal solution is found.

6.3.4.5 Example Let us consider the following maximization problem:

$$\text{maximize}\quad 3x_1 + 2x_2$$

subject to

$$2x_1 + x_2 \leq 4$$
$$x_1 + x_2 \leq 3$$
$$x_1, x_2 \geq 0$$

This problem can be converted into a minimization problem with the addition of slack variables:

$$\text{minimize}\quad -3x_1 - 2x_2 + 0x_3 + 0x_4$$

subject to

$$2x_1 + x_2 + x_3 = 4$$
$$x_1 + x_2 + x_4 = 3$$
$$x_1, x_2, x_3, x_4 \geq 0$$

Then, what is called the *tableau* can be constructed.

2	1	1	0	4
1	1	0	1	3
−3	−2	0	0	0

From the last row, it can be seen that the most negative element of the objective row is −3 which indicates that the first column is the one that can be included into the column. To leave the basis, the first row indicates that the value is $\frac{4}{2} = 2$ and the second row gives $\frac{3}{1}$ which means that column 3 leaves the basis. Solving the system again gives the following tableau:

1	$\frac{1}{2}$	$\frac{1}{2}$	0	2
0	$\frac{1}{2}$	$\frac{-1}{2}$	1	1
0	$\frac{-1}{2}$	$\frac{-1}{2}$	0	6

The most negative value indicates that column 2 enters the basis and then the row calculation says that column 4 is leaving it. Solving it again gives

1	0	1	−1	1
0	1	−1	2	2
0	0	1	1	7

Since there is no negative coefficient in the row corresponding to the cost function, the optimal value, which is $(1, 2)$ is obtained.

6.4 OPF APPLIED TO COMPETITIVE ELECTRICITY MARKETS

In a competitive electricity market, an important function of the market operator is to determine the prices of the generator energy sales and the associated transmission services, requiring the calculation of spot or real-time prices. The nodal prices of generation buses are also discovered by solving the economic dispatch problem. An extension of this formulation to price for all buses is possible within the framework of an OPF in which the objective function is the minimization of the total system production cost.

The Lagrangian function at the optimum $(\mathbf{x}^*, \mathbf{u}^*, \mathbf{p})$ is defined as

$$\pounds = f(\mathbf{x}^*, \mathbf{u}^*, \mathbf{p}) + \lambda^t \mathbf{g}(\mathbf{x}^*, \mathbf{u}^*, \mathbf{p}) + \mu^t \mathbf{h}(\mathbf{x}^*, \mathbf{u}^*, \mathbf{p}) \qquad (6.19)$$

The derivative of this function with respect to P_i is

$$\frac{\partial \pounds}{\partial P_i} = \lambda_i \qquad (6.20)$$

Therefore, this equation quantifies the cost variation at the optimal point due to generator or demand power incremental change at each bus. Depending on the method used to solve the OPF, this value can be obtained directly or indirectly. For example, using the linear programming approach, the Lagrangian function at the optimum point is

$$\pounds = f(\mathbf{x}^*, \mathbf{u}^*, \mathbf{p}) + \lambda^t \mathbf{G}(\mathbf{x}^*, \mathbf{u}^*, \mathbf{p}) \qquad (6.21)$$

and since the gradient is zero at the optimal point

$$\nabla \pounds_x = \frac{\partial \mathbf{f}}{\partial \mathbf{x}} + \left[\frac{\partial \mathbf{G}}{\partial \mathbf{x}} \right]^t \lambda = 0 \qquad (6.22)$$

which gives

$$\left[\frac{\partial \mathbf{G}}{\partial \mathbf{x}}\right]^t \lambda = -\frac{\partial \mathbf{f}}{\partial \mathbf{x}} \tag{6.23}$$

The coefficient matrix has N rows (number of state variables) and M columns (power balance constraints), where $N \leq M$. This system can be solved using the least square techniques.

6.5 FURTHER DISCUSSIONS ABOUT OPF CALCULATIONS

Traditionally, OPF terminology has been differentiated into two broad terms: (1) OPF, which is the optimization problem applied to the base case and (2) security-constrained optimal power flow (SCOPF), which specifies the optimization process that includes additional security constraints including contingencies. Using these definitions, OPF becomes a special case of SCOPF. In general, practical power systems inevitably include security constraints. Therefore, inclusion of security constraints in the OPF problem significantly influences the selection of the solution method. However, this fact was not properly considered by the researchers over the years, and hence further research on this topic is definitely needed. New advances in computing power, mathematical programming methodologies and their solvers make it possible to develop SCOPF tools accordingly to meet the needs of large-scale transmission system operators and market operators.

In addition, transmission system operators face multiple and often competing objectives which typically include the cost minimization, emission minimization, and loss minimization. Therefore, mathematical formulations that can efficiently handle multiobjective functions are needed. Moreover, multiperiod OPF problems need to be further formulated because the OPF solution of one optimization period (say a market-clearing interval) affects the OPF problem and its solution of the next period. The continuous evolution of market prices, the more volatile behavior of loads, generation ramping units, the inclusion of storage devices, and limitations of transformer tap changing over the periods—all these factors make it necessary to solve the optimization problem globally over several time periods. Another essential control option to be investigated is the demand response management, which brings additional degrees of freedom with respect to load schedule. This type of model also requires extending the optimization over a longer time horizon leading to multiperiod OPF applications.

To complicate matters further, there is a continuous development of new and sophisticated devices used in power systems such as phase angle regulator, *high-voltage direct current* (HVDC), and *flexible alternating current transmission system* (FACTS) devices which are mainly used for power control. It is both important and imperative to have an accurate modeling of these devices within an OPF formulation to achieve optimal solutions which also represent the operating characteristics and limits of these devices.

6.6 ALGEBRAIC MODELING LANGUAGES AND SOLVERS

Any of the optimization problems presented in this chapter requires a concrete model data structure to include it into a solver. In addition, they need a good way to extract the final results for further analysis and for making decisions. That is the reason for the significant development of modeling languages and software applied to large-scale optimization problems. While there are many modeling languages currently available, some of the notables include GAMS, AIMMS, and AMPL, among the commercially available modeling language software, and Pyomo, Pulp, CyLP, and yoposib among the open-source softwares. Once a particular model is set up, different solvers can be selected, depending on the type of problem. Commercially available solvers are CPLEX, Gurobi, XPRESS, MoSEK, and LINDO. Open-source software solvers are GLPK, CLP, DYLP, etc.

CHAPTER END PROBLEMS

6.1   Find the feasible region and the extreme points of the following linear programming problem.
Maximize

$$z = -4x_1 + 7x_2$$

subject to

$$x_1 + x_2 \geq 3$$
$$-x_1 + x_2 \leq 3$$
$$2x_1 + x_2 \leq 8$$
$$x_1, x_2 \geq 0$$

Obtain the solution graphically.

6.2   Find the solution of the previous item using simplex method.
Maximize

$$z = -2x_1 + 3x_2$$

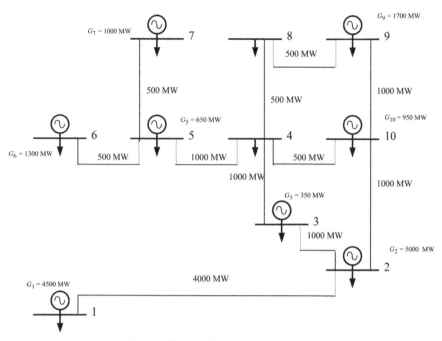

Figure 6.7   Ten Bus Equivalent System.

subject to

$$x_1 + x_2 \geq 3$$
$$5x_1 + 7x_2 \leq 35$$
$$4x_1 + 9x_2 \leq 36$$
$$x_1, x_2 \geq 0, \in N$$

6.3 Does every linear problem have an optimal solution?

6.4 Consider the equivalent 10-bus system illustrated in Figure 6.7, and the parameters described in Tables 6.9 and 6.10. Voltage limits are set to +/− 5%, and branch limits are described in the Figure. Solve the OPF related to the system.

6.5 Find the feasible region and the extreme points of the following linear programming problem.
Maximize

$$z = -4x_1 + 7x_2$$

subject to

Table 6.9 Nodal Load

Node	P (MW)	Q (MVAr)
1	300	100
2	5.000	2.000
3	50	10
4	700	140
5	500	140
6	200	60
7	300	90
8	300	110
9	1.000	300
10	200	60

Table 6.10 Generation Cost

Generator ID	a	b	c	P^{Min}	P^{Max}
1	0.0	1.0	0.003	0.0	5.800
2	0.0	9.5	0.002	0.0	5.000
3	0.0	1.0	0.050	300	350
5	0.0	10.0	0.050	600	650
6	0.0	9.0	0.010	0.0	1.300
7	0.0	7.1	0.040	0.0	1.000
9	0.0	1.0	0.001	0.0	1.700
10	0.0	1.0	0.001	900	950

$$x_1 + x_2 \geq 3$$
$$-x_1 + x_2 \leq 3$$
$$2x_1 + x_2 \leq 8$$
$$x_1, x_2 \geq 0$$

Obtain the solution graphically.

6.6 Find the solution of the previous problem using simplex method.
Maximize

$$z = -2x_1 + 3x_2$$

subject to

$$x_1 + x_2 \geq 3$$
$$5x_1 + 7x_2 \leq 35$$
$$4x_1 + 9x_2 \leq 36$$
$$x_1, x_2 \geq 0, \in N$$

FURTHER READING

1. Carpentier J. Contribution à l'étude du dispatching économique. *Bulletin de la Société Française des Électriciens* 1962;Ser.8(3):431–447.

2. Dommel HW, Tinney WF. Optimal power flow solutions. *IEEE Transactions on Power Apparatus and Systems* 1968;PAS-87(10):1866–1876.

3. Sun DI, Ashley B, Brewer B, Hughes A, Tinney WF. Optimal power flow by Newton approach. *IEEE Transactions on Power Apparatus and Systems* 1984;PAS-103(10):2864–2880.

4. Stott B, Marinho JL. Linear programming for power system network security applications. *IEEE Transactions on Power Apparatus and Systems* 1979;PAS-98(3):837–848.

5. Alsaç O, Bright J, Praise M, Stott B. Further developments in LP-based optimal power flow. *IEEE Transactions on Power Systems* 1990;5(3):697–711.

6. Wu YC, Debs AS, Marsten RE. A direct nonlinear predictor-corrector primal-dual interior point algorithm for optimal power flows. *IEEE Transactions on Power Systems* 1994;9(2):876–883.

7. Pedregal P. *Introduction to Optimization*. New York: SpringerVerlag; 2004.

8. Gass SI. *Linear Programming: Methods and Applications*, 4th edition. McGraw-Hill; 1975.

9. Baldick R. *Applied Optimization: Formulation and Algorithms for Engineering Systems*, 1st edition. Cambridge University Press; 2009.

10. Fisher ML. The Lagrangian relaxation method for solving integer programming problems. *Management Science* 1981;27(1):1–18.

11. Land AH, Doig AG. An automatic method of solving discrete programming problems. *Econometrica* 1960;28(3):497–520.

12. Marchand H, Martin A, Weismantel R, Wolsey L. Cutting planes in integer and mixed integer programming. *Discrete Applied Mathematics* 2002;123(1–3):397–446.

13. Jensen PA, Bard JF. *Operations Research Models and Methods*, 1st edition. Wiley; 2003.

Chapter 7

Design, Structure, and Operation of an Electricity Market

This chapter will cover the basic designs, structures, and operations of electricity markets. The materials covered in this chapter were largely drawn from the characteristics of the electricity markets that are currently operating in the United States, certain Latin American countries, and some European countries. There is no single standard design in existing electricity markets as there are many variations in such designs and structures. Comparison among these designs and structures is also a debatable issue because each design and structure has its own merits and disadvantages. Market design is the primary means of achieving and promoting competitive outcomes in electricity markets. Proper market design will promote competition by laying the foundation for a structural basis for a competitive outcome. Other measures must be taken in certain instances in which the market structure is not competitive. Both structural and non-structural changes will be necessary to ensure that electricity markets are competitive.

7.1 ELECTRICITY MARKET OVERVIEW

Electricity market is one of the key outcomes of electric industry restructuring which was extensively treated in Chapter 1. The intent to open up the generation sector and expose it into competition led to the development of many kinds of electricity markets. Generally, an electricity market is a way or mechanism to exchange electricity among different participants, entities, or different countries. It is also designed to

Electricity Markets: Theories and Applications, First Edition. Jeremy Lin and Fernando H. Magnago.
© 2017 by The Institute of Electrical and Electronics Engineers, Inc. Published 2017 by John Wiley & Sons, Inc.

provide incentives for building generating capacity or consuming electricity where appropriate. Economists believe that markets are one of the best ways to trade any commodity including electricity. More formally, an electricity market can be defined as a centralized mechanism through which participants can exchange electricity transparently according to the price they are willing to pay or receive, subject to the capacity of the electrical network. Electricity market is a general term that represents a collection of various kinds of markets in which electricity is traded for various system requirements. System requirements include capacity, energy, regulation, and ancillary services. At least in the United States, the electricity markets were developed on a voluntary basis. Some of the objectives of the designed markets include (1) shifting risk from rate-payers (electricity consumers) to the generating firms which will profit from power markets, (2) attracting investment capital so that the retiring power plants can be replaced, and (3) ensuring that new types of generating resources can participate in the markets.

Also, an electricity market operates within a clearly defined boundary. This boundary is generally defined for a large region (balancing authority) enclosed by its electric network. In most cases, the electrical network boundary within which the power system operates also defines the boundary for a particular electricity market. From the economic standpoint, the clear definition of a market boundary is extremely important because that defined boundary will determine which elements or participants will or will not be part of a particular market. For example, all generators that are physically located within a particular market boundary should be considered as part of the overall supply curve for that market. By the same token, the import power into a market system must also be considered as part of the supply curve for the same market. Those generators that are physically located outside of the defined network boundary but are eligible to participate in the relevant market can also be considered as part of the supply curve. Therefore, defining a market boundary is a critical first step.

Generally, an electricity market can be broadly classified into two different types: *energy market* and *capacity market*. Both markets are complementary and closely related. This classification was originated from the same concepts used in the power system field. The capacity of a generating resource is the maximum capability of the resource in providing energy into the power system. The actual energy produced in real-time operation by that resource may or may not reach the level of its maximum capacity. Thus, the energy market concerns the actual energy produced by the participating generation resources while the capacity market relates to the availability of the maximum capability of the same resources. This distinction is extremely important to understand these two different markets. Some market designs include energy market only. These markets are called *energy-only markets*. Some other designs have both energy and capacity markets. We can call those markets *energy-capacity markets*.

An energy market provides a venue or mechanism in which power demand can be matched with power supply with consideration of other technical constraints. The energy market is typically for near term such as day-ahead, hour-ahead, or real-time. Market clearing of an energy market for a longer time horizon such as a week-ahead, is quite challenging because it will require forecast load and future offers/bids by the suppliers/purchasers which are significantly uncertain. Energy markets provide market-clearing prices which become the basis for determining earnings by generators, payment by load, and hedging of congestion.

As a complement to the energy market, a capacity market is designed to provide additional revenues for generators which may not earn sufficient revenues in energy markets alone. Not all electricity markets have a capacity market. And not all capacity markets that exist today have the same features. Capacity markets are also evolving. It is still a question of whether an energy-only market provides sufficient revenues to generators for their long-term economic viability. Also, it is still a debate whether the capacity market, in combination with its energy market, brings reliability and other tangible benefits to the market system as it was designed to.

Among the energy markets, day-ahead and real-time (balancing) markets are predominant. For example, most of the electricity markets in the United States and European countries have both day-ahead and real-time markets although the actual markets are cleared quite differently. Some electricity markets have intraday energy markets in which markets are cleared an hour or half-an-hour ahead of real-time system operation. Generally, the day-ahead market is a forward market while the real-time market is a spot market. However, day-ahead market is sometimes treated as part of a spot market.

Generally, there are two pricing schemes in any electricity market: *nodal pricing* and *zonal pricing*. Electricity markets in the United States primarily use nodal pricing scheme while those in Europe predominantly employ zonal pricing. Nodal pricing is also called locational marginal pricing (LMP). These pricing schemes will be treated in more details in a later chapter. One of the objectives of the market pricing scheme is to provide both short-term and long-term price signals to market participants so that market participants can make appropriate economic decisions in a market setting. The short-term price signals are typically provided by zonal/nodal prices from day-ahead, intraday, and real-time markets. The long-term price signals are provided by a longer-term capacity auction market or in some cases by electricity futures market. In other words, energy-capacity markets are generally better than energy-only markets in providing longer-term price signals. Those prices reflect the results of a complex combination of economic decisions made by market participants, through offers and bids, based on the expected and/or actual system network topology.

7.2 ELECTRICITY MARKET MODELS

There are many possible ways or schemes that are currently used to trade or contract electricity among sellers and buyers. The most common forms of electricity trading are bilateral contracts, power pool model, power exchange model, or some sort of combination. Self-schedule of electricity is not included in the electricity trading because in a self-schedule, an entity is using its own generating resources to meet its own load. Thus, it is not the same as electricity trading that we discuss here. Also, the discussion of the internal pricing used in a self-schedule arrangement, known as *transfer pricing*, is beyond the scope of this book. Bilateral contracts are based on the results of negotiations that are carried out directly between the parties (buyers and sellers) involved without a central clearing house. The bilateral contracts are known as *non-organized* or *over-the-counter* markets as parties have to agree on the terms and conditions of the transactions among each other without the support of any central entity.

Power pool, power exchange, and later ISO models are known as *organized markets* as these markets are cleared by a central market-clearing entity. These models are the primary models of electricity markets discussed in this book. In such markets, sellers provide *offers* and buyers submit *bids* to a central market-clearing entity. Then, this central entity matches the supply offers with demand bids and determines the final winning sellers with winning MW schedules using some sort of auction mechanisms. The prevailing form of auction used in these market-clearing mechanisms is uniform-price auction. Market prices are also determined from these auctions.

These organized markets are structured markets in which the trading procedures and the structural conditions to operate are established based on the defined market rules. The set of market prices (both forward and spot market prices) that is published by the organized markets represents an important reference for the financial and bilateral trades that take place in parallel markets. Both organized and non-organized markets are complementary.

An entity which is responsible for the management of the organized markets (either power pool or power exchange model) is called *electricity market operator* or simply *market operator*. Occasionally, the term "market operator" may be used to represent the power exchange. The primary functions of a market operator are to assure and promote the market competitiveness, as well as to perform financial settlements for the consummated transactions. The market operator complemented by the energy regulator, should guarantee the integrity and good governance of the electricity market.

7.2.1 Power Pool Model

Power pool models were prevalent in both the United States and the United Kingdom even though there were significant transformations to these models. Each of the models is further explained below. It is important to understand power pool model as it is one of the key models of electricity markets.

In the United States, power pools exist well before the existence of electricity markets. For example, PJM Power Pool was established in 1927 with just three electric utilities. The primary goal of establishing a power pool is to improve the economic efficiency and increase the economic benefits made possible by interconnecting two or more electric utilities, sharing their generating resources, and jointly managing the power flow. The economic benefit of lower dispatch costs (savings) was shared among the transacting parties. In 1993, PJM Power Pool began the transition to an independent, neutral organization when the PJM Interconnection Association was formed to administer the power pool. In 1997, PJM became a fully independent organization. At that time, membership was opened to non-utilities and an independent Board of Managers (aka Board of Directors) was elected. The new organization serves more than its original pool members.

As another example, New York Power Pool (NYPP) was formed in 1969 in response to the first Northeast Blackout, which occurred in 1965. NYPP was established as an overseeing organization to prevent another incidence of a widespread blackout. After the pool was established, however, the pool utilities were entering into more and more bilateral agreements of power sales with each other and with other entities beyond their pool boundary. Later, pool members pursued activities which were beyond the original objective of forming a power pool by signing these contracts without regard to the benefits of shared savings. During the same periods, there was a growing number of independent generators, known as independent power producers (IPPs). These entities were originally spawned by legislation guaranteeing them a price and a captive market for their generated power. Later, some of the major utilities from the pool started transacting with IPPs as well. The IPP generators were committed and dispatched by their respective owners and only responded to NYPP directives under emergency conditions. With so many energy transactions taking place without a central coordinating facility, it was difficult to decide which transactions to curtail in the event of a failure of a transmission line or a generator. These transaction curtailments must also be made quickly in order to prevent further problems caused by overloads on the remaining lines.

Finally, FERC stepped in and took series of actions by encouraging areas such as New York, New England, and California, to form independent entities that would manage the transmission system. It was later concluded that NYPP, comprised of eight member utilities in the state of New York, would not be neutral and would be under pressure to protect their own interests rather than operate the transmission system in the most efficient and fair manner for the benefit of every market participant. The ISO model rather than Power Pool model appears to fit that role which would be a neutral entity that would not have any financial interest in the outcome of the market.

The ISO model, as was being proposed for New York State and that would be empowered by FERC, was to form an independent, not-for-profit company which will serve two primary functions: (1) to operate the transmission system fairly and efficiently and (2) to act as a broker to purchase the power from generators and sell it to the *load-serving entities* (LSE) to meet their load. The second function is similar to the key function provided by a power exchange. In other words, the dual role of the ISO would be to manage the transmission system reliably and to operate the competitive electricity market.

The transition from power pool model to ISO model which operates the nodal electricity market naturally made the association between power pool and nodal pricing very close. During this transition, the ISO model-based market operators quickly learned that zonal pricing missed to capture the internal congestion that would have occurred. Therefore, all the electricity markets in the United States quickly adopted the nodal pricing scheme which properly captures the internal congestion of the market system.

In an ISO market model, generators offer and demand bid into the market. Then, the market is cleared by matching supply and demand with consideration of other technical constraints, such as transmission limits and generator limits. Nodal prices at various nodes throughout the system are computed, along with scheduled generation for selected (committed) generators. During market clearing, these markets also make sure that the committed generators would not lose money, hence *make-whole payments* or *uplifts* are incorporated when committing these generators.

The development of power pool model in the United Kingdom, particularly in England and Wales took a different path. The power pool model, used in the literature primarily refers to the power pool model of the United Kingdom, which was operational from 1990 to 2001. Even that UK power pool model has transformed into a different model after the introduction of New Electricity Trading Arrangements (NETA) and the British Electricity Trading and Transmission Arrangements (BETTA). The Electricity Act of 1989 laid the legislative foundations for the restructuring and privatization of the electricity industry in Great Britain. The act made provision for a change in ownership from the state to private investors, the introduction of competitive markets, and a system of independent regulations. The electricity industry was restructured *prior* to its privatization. As part of reform in England and Wales, the electricity pool was established as the wholesale market mechanism through which electricity was traded among buyers and sellers. The pool was one of the first mechanisms of its kind as an electricity market model.

In its development, considerable weight was given to the arrangements operated pre-privatization by the CEGB, when the electricity system was publicly owned and centrally planned. The principles of the pool were relatively simple and largely inherited from the CEGB merit order. The pool's role was to (1) develop a set of rules defining how electricity in the market was to be traded, (2) develop a market system through which generators had to offer wholesale electricity and from which those who wanted to purchase wholesale electricity had to buy, (3) set wholesale electricity prices for each half hour, and plants were dispatched, and (4) conduct financial settlements, by which generators were paid and load suppliers were charged.

The pool was set up to facilitate a competitive bidding process between generators that set the price paid for electricity each half hour of the day and established which set of generators would run to meet the forecast demand. The pool was an unincorporated association of its members. Its members, wholesale buyers and sellers of electricity, controlled how the pool was run and decided if and how the pool should change. The National Grid Company (NGC) operated the pool and administered the pool's settlement system on behalf of pool members.

The pool required generators, each day on a day-ahead basis, to provide details of the price at which they were prepared to make generation available. NGC, on behalf of the pool members, provided an estimate of system demand at the day-ahead, calculated a schedule of generation to meet this estimate, and determined pool prices. NGC was also responsible, under its license, for dispatching plant on the day, taking into account the day-ahead schedule, but modifying it as necessary, for example, to take into account of unexpected changes in demand or failures by generating plants and to resolve system constraints.

One can argue that the major players in a power pool are generators (sellers) who largely controlled the wholesale prices of electricity because participation by the demand is quite limited. Power pool system in the United Kingdom survived from 1990 through 2001. Since 2001, they adopted a new trading system called NETA which replaced the previous power pool model. In NETA, both generators and demand sides can set the wholesale price of electricity. In 2005, another new trading system, BETTA was introduced replacing NETA system. The objective of creating BETTA was to create, for the first time, a fully competitive British-wide market for the trading of electricity generation (the wholesale market) by extending NETA into Scottish market. The transmission system in England and Wales will still be operated by NGC. In Scotland, a separate system is run by ScottishPower and Scottish and Southern Energy. The two systems are joined by an England/Scotland Interconnector. This is to ensure the open and fair access to the British-wide transmission network.

The application of nodal pricing is more common in a power pool but zonal pricing is entirely possible. Regardless of the pricing model, the objective function of a power pool is to determine the least-bid dispatch with resultant market prices. The key distinction between zonal and nodal pricing is that transmission network constraints internal to the system are explicitly modeled in nodal pricing while those constraints are ignored in zonal pricing.

7.2.2 Power Exchange Model

Power exchange (PX) or simply *exchange* is one of the two prominent models of organized electricity markets. In fact, PX is the predominant form of electricity trading in European markets. It is a place where electricity can be traded for physical consumption or for purely financial transactions or both. The power exchanges represent a centralized entity in which the supply and demand meet to create a competitive market in which the electricity is procured. The market clearing in power exchanges is done by simply matching supply offers with demand bids without considering the network constraints. PX can support electricity trading for next day as in a day-ahead market or for next hour or next half-an-hour as in an intraday market.

Exchange typically uses an auction mechanism to match between bids by buyers and offers by sellers. Auction markets are one type of exchange market models. An exchange acts as a counter party to all trades that occur in an exchange. In other words, an exchange entity acts as a buyer for a selling party and acts as a seller to a buying party for trading electricity. This eliminates the burden of assessing the credit worthiness of the counter party.

Advantages of a power exchange over a bilateral market include (1) reduced trading costs, (2) increased competition, (3) market prices that are visible to buyers/sellers/traders, and (4) faster operation due to standardized contracts. Some disadvantages include (1) facilitate collusion and (2) less flexiblility for contracts because typically the size, terms, and conditions of the contracts in an exchange are standardized.

Other benefits of PXs include easy access, low transactions costs, neutral market place and price reference, safe counter party and clearing, and settlement service. PX is also helpful for a well-functioning of the retail market since a liquid PX gives the retail suppliers an opportunity to procure energy without the need to own the production capacity.

Exchanges can operate faster than bilateral markets so they can operate much nearer to real-time than bilateral markets. Exchange models use simple bids and relies on a single price which clears the market. Therefore, there are no side payments to make-whole the generators when offering into a power exchange. The generators do not take account of their start-up costs and no-load costs. Power exchanges do not allow generators to include those costs. However, bid manipulation by generators to account for these costs are a reality. Different power exchanges can be set up for different system requirements, such as incremental energy, decremental energy, reserve, ramp, and load-following.

The main goals of the PXs are to facilitate the electricity trading in a short term, and the promotion of information, competition, and liquidity. PXs differ from country to country with respect to the diverse market design, regulatory framework, and the background of the electricity industry. Notable power exchanges in European markets include Nord Pool (Nordic and Baltic region), APX (the United Kingdom, Netherlands, and Belgium), EPEX (France, Germany, Austria, and Switzerland), and Omel (Spain).

7.3 ENERGY MARKETS

One of the primary objectives of the electricity market operation is to facilitate the economic functioning of the electrical system while ensuring a secure and reliable network grid operation. Along with secure grid operation, one of the main tasks of the electricity market is to calculate market prices which provide economic signals to expand generators and industrial loads, optimize the use of generation capacity and development of renewable energies.

The electricity market operation rules need to be established to guarantee the fair access to the network and to promote the efficient investment in the market assuming that this effect reduces the cost of electricity. The market operation must maintain the efficient operation and use of all electricity services. This needs to be done mainly for the long term interests of electricity consumers and the signals need to reflect the market efficiency primarily on the price signal but also to maintain and improve the quality, safety, reliability, and the security of electricity supply.

As was mentioned before, one of the unique aspects of the electricity markets compared with other commodity markets is that electricity needs to be generated, distributed and delivered in real-time. Therefore, although the primary objective of an electricity market is to calculate the market prices, its complexity resides in the real-time operation of the electrical system. It is important to remark that secure operation must be ensured at all times regardless of how the market is operated, since system security is the most important aspect of the grid operation.

Market operation models, data requirements, and rules are different for different electricity markets. These requirements depend on the specific characteristics and conditions of each market. Some of these attributes include types of energy resources and demand, network grid characteristics, public policies, and weather conditions. In this chapter, general features of both energy and non-energy markets are presented to illustrate the central aspect of the market operation without losing generality. To date, many electricity markets have already been established around the world. However, there is a steady evolution of policy objectives for different systems, in addition to the several developments in other fronts. Some developments comprise of the adoption of new technologies, the increasing need for private investment, and promotion of competition and efficiency. Furthermore, it is necessary to increase the utilization of non-fossil fuel sources of energy so as to diversify the generation mix as well as to increase the flexibility in system operation. All these developments make it imperative to reconsider and reevaluate existing electricity market designs in an effort to meet a broader range of objectives and constraints. Therefore, it is crucial to know models and techniques applied to electricity markets.

Energy markets generally include day-ahead market, real-time market, and intra-day markets (an hour or half-an-hour ahead). Day-ahead and real-time markets are predominant in many markets while some markets have intraday markets as well.

7.3.1 Day-Ahead Energy Market

Among the energy markets, the day-ahead energy market or day-ahead market is one of the most important markets. Day-ahead market is a forward market in which hourly clearing prices (either zonal or nodal prices) are calculated for each zone/node in the system, for each hour of the next operating day. The computation of market-clearing prices is based on generation offers, demand bids, virtual (financial) bids in some markets, and bilateral transaction schedules submitted into the day-ahead market. Depending on the bidding rules of the market, the bids can be simple bids, block bids, and/or complex bids. In the nodal pricing, market-clearing algorithm also uses the expected network topologies for each hour of the next operating day.

Conceptually, there can be as many as 24 different network topologies to be used in the day-ahead market clearing, assuming there is a change in network topology for each hour of next day. Regardless of the network status for next day, the day-ahead market is essentially a set of 24 constrained optimization problems to be solved given the input data provided to the market. The solutions from the day-ahead market clearing for next day represent a set of solutions of each hourly optimization problem for next operating day. The day-ahead market solutions include generation schedules (MW) for winning generators (including virtual bidders in some markets) to supply to the market, consumption schedules (MW) for demand consumers to consume and demand responders to forgo consumption, market prices for each relevant zone/node in the system and congestion (rents and binding hours) related to binding transmission facilities for each hour of next day.

In terms of the bidding schedule, market participants, generators, loads, and financial players have to submit their offers and bids for the next operating day by certain deadline today. For electricity markets in the United States, this bid-closing time varies from market to market. For PJM, the bid-closing time is currently set at 10:30 a.m. (EPT – Eastern Prevailing Time). The bid-closing time for MISO day-ahead market occurs at the same time of 10:30 a.m. (EPT). After that period, the bidding for the day-ahead market is closed. During the next several hours, the electricity market operator will run the market-clearing engine, which is the process of finding the optimal solutions to the constrained optimization problems mentioned previously. Results for day-ahead markets in both PJM and MISO are posted at 13:30 p.m. (EPT) on daily basis. In the zonal markets, there will be a single market-clearing price for the entire zonal market as network models and constraints are typically not considered.

During the market-clearing process in the nodal markets, the day-ahead schedules of next-day generation dispatch and market prices are developed using the least-bid security-constrained unit commitment (SCUC) and security-constrained economic dispatch (SCED) algorithms based on the offers/bids. The objective of this algorithm is to minimize the total bid-based production cost of the system subject to various system constraints. System constraints typically include generators' operational constraints, transmission constraints, and other constraints. Once market solutions are found, the market operator posts the day-ahead market prices and sends hourly energy schedules for selected (committed) units, for the next operating day. The day-ahead market results—energy schedules and associated market prices—are contractually and financially binding to all market participants. The committed generators will be paid for their scheduled generation at market prices at their respective generator buses or zones, while the load customers will pay for their scheduled consumption at market prices at their respective load buses or zones. Prices from the day-ahead market also provide guidance to contract prices of bilateral transactions among buyers and sellers of electricity.

Day-ahead market can be complemented with several intraday markets, such as hour-ahead market in which each participant can adjust their open positions up to a point closer to the real-time. The participants can diminish their out-of-balance risks, to avoid the payment for imbalance prices. This can help lower the balancing load on grid operators. Some markets allow rebidding for units that do not get selected in the day-ahead market clearing. So, these units can rebid their offers for possible participation in next day real-time market.

7.3.1.1 Day-Ahead Market Operation

Once the input data are decided, appropriate computational tools for market clearing must be selected. The choice of these tools will depend on the market time frame: real-time, day-ahead, or longer simulation time. The analytical tools mostly used by the market operators should be able to do the reliability unit commitment, the security-constrained unit commitment, the security-constrained economic dispatch and the feasibility tests. While the day-ahead market determines hourly market-clearing prices and unit commitments, the market operator also needs to analyze the needs for unit must-run, resolve uplift issues and mitigate bids if necessary with the goal of producing the least-bid energy schedule while meeting the reliability needs.

The day-ahead market clears after receiving the following input data: the network model, the approved transmission outages, the physical schedules, the demand bids, the supply offers and resource definitions. The three key processes in the day-ahead market are (1) market power mitigation, (2) integrated forward market, and (3) residual unit commitment. If any bid fails the market power test, it is mitigated typically immediately by the tool by following the market rules. And the system determines the minimal and most efficient schedule of generation to address local reliability.

Day-ahead market-clearing tool uses an SCUC, SCED, and then a simultaneous feasibility test (SFT). These steps are used to clear the supply offers and demand bids while taking into account of the constraints, and the locational marginal prices are determined. In some markets, once the results are posted by the market operator, the market participants can adjust the schedules and submit them again to the market operator. Those market participants should not be those who are selected in day-ahead market clearing. Then, in addition to the day-ahead market schedules and real-time load forecast, additional information such as start-up and no load offers and the resource definitions are received. In some markets, start-up and no load offers are already considered in the day-ahead market clearing. With additional information, a reliability assessment commitment (RSC) using a security-constrained unit commitment algorithm is performed. The results from this step finally establish the resource commitment for the next day.

The integrated forward market simultaneously analyzes the energy and ancillary service markets to determine if there are congestions in the transmission network, which is also known as *congestion management*, and confirms the reserves required to balance supply and demand based on supply offers and demand bids. It ensures that the sum of generation and imports equals the sum of load, exports, and transmission losses. It also ensures that the final schedules are feasible with respect to the constraints enforced in the full network model as well as all the ancillary service requirements. When forecasted load is not met in the integrated forward market, the residual unit commitment process enables the market operator to procure additional capacity by identifying the available least-cost resources not committed earlier.

7.3.1.2 Input Parameters for Market Clearing The market-clearing model requires different groups of input parameters to find optimal dispatch schedule and market prices. The most important required input data are (1) electricity demand, (2) generating plant parameters, (3) transmission network parameters, and (4) various restrictions and constraints.

7.3.1.3 Demand Parameters Demand for electricity is defined as an exogenous variable on hourly, daily, and yearly basis. It includes net electricity consumption of end-use consumers, transmission losses within model regions, and other conversion losses in the electric grids. On the other hand, consumption for the storage operation, cross-border transmission losses, and the power plants' consumption are modeled as endogenous variables. The installed capacity needs to ensure enough level of backup power to ensure the security of supply.

The fixed demand is determined based on the forecast. The demand bids can be defined in different ways. The first one can be an aggregated load. The market participants submit the fixed demand for the periods that they wish to purchase from the specific market. Required parameters for demand bids include MW quantity, location, and period. The addition of fixed demand shifts the aggregate demand curve to the right. In addition to the fixed demand bid, the price-responsive demand (PRD) can be specified. Through PRD, the market participants propose the demand (and energy) that they are willing to buy at a specific price. They are willing to forgo their consumption if the market price exceeds their willingness-to-pay price. This price-responsive demand is calculated using distribution factors for all load buses in the system to break down demand bids across the load zones. These distribution factors are calculated by the market operator. The three parameters required for submitting bids for PRD are MW/price pair, location, and period.

7.3.1.4 Generator Parameters Generation technologies are defined by their respective economic and technical input parameters. These parameters include costs, restrictions, and emission factors. Some of these parameters are described below.

The energy bids or costs include three main components.

- **Energy Bid Cost**: The generation bids can be submitted in aggregated form associated with generation, if a unit is on cost-based and the production cost curve is based on fuel costs and efficiency rates. The generator cost curve must be monotonically increasing.
- **Start-Up Cost**: This is the cost associated with a situation when a unit status is changed from off-line to online status. In general, this information is specified with three parameters instead of a curve: hot, intermediate, and cold start-up costs. In addition, generators need to submit the transition times between these statuses.
- **Minimum Load Cost**: The minimum load cost or no load cost expresses the unit operating cost at the minimum operating point. The minimum load cost is considered whenever a generating unit is online.

These components of energy bids or costs are included in the objective function. Additionally, generators need to submit several parameters that are included in the constraint set:

- **Status and Capacity Changes**: Normally, the possible unit statuses are (1) unavailable either for commitment or dispatch, (2) economic which means that the unit is available for commitment and dispatch, (3) emergency which means that the unit is only available during emergency, (4) must-run meaning the unit is committed by the market participant request, and (5) regulation self-schedule and reserve self-schedule.

- **Operational Constraints**: A unit's mechanical or economic upper and lower production levels. Normally two levels are required: dispatchable and emergency minimum and maximum. Non-dispatchable units need to set dispatchable minimum and maximum at the same values.

- **Temperature Sensitivity Limits**. These temperature values are used to derate the generator dispatch limits based on the current ambient temperature.

- **Energy Production Capacity**: This constraint establishes the maximum capacity (MWh) for the unit within the simulation time frame. For example, this energy constraint can simulate the fuel availability of a thermal unit.

- **Ramp Rates**: Ramp rates within the range of production levels represent the time required for the generator to move from one production level to another while respecting the turbine's safe thermal gradients.

- **Minimum Run and Minimum Down Times**. The commitment schedules are decided for consecutive hours that are equal to or greater than the minimum run time. The minimum down time means the minimum time required for the generator to stay offline once the unit is set offline from the system. The market will have commitment schedules that do not violate the minimum down time.

- **Maintenance Schedules**: This is also known as must-off, which determines the periods that the unit must remain off due to maintenance.

- **Operating Reserves**: This is the capacity between unit current power output and the maximum limit. This operating reserve allows the unit to quickly increase the generation in the event of a generator or transmission outage or as necessary.

- **Must-Run Status Contracts or Other Requirements**: A unit will be assigned a must-run or must-take status so that it is not fully dispatchable.

- **Limits and Costs of Emission and Control Allowance**: For example, units that use up their emission allowances prematurely may not be available to operate during peak periods.

- **Maximum Daily and Weekly Starts**: It is the maximum number of times a unit can be started in one day or one week.

Information requirements of some of these parameters differ among different markets: real-time, day-ahead, or longer term markets. For example, a unit's availability on the date and time in question might be affected by the following factors:

1. Inclement weather, prior performance problems, or fuel availability
2. Minimum sustained production levels to keep the unit available for the next hour or next day
3. A unit's ability and contractual requirement to deliver ancillary services, such as reactive power or quick-start capability
4. A forecast of expected unit production levels at different points in the dispatch period for hydro, wind, or other intermittent resources

The aggregated generation bids are optimized in aggregated form and resulting generation schedules are disaggregated into the schedules of individual generating units using *generation distribution factors* (GDF) to perform power flow calculations.

7.3.1.5 Transmission Network Parameters and Restrictions

Transmission network parameters are needed to incorporate the electrical system model into the mathematical representation of the scheduling problem. The network model ensures that the flows between generation and loads are physically feasible, and there is no limit violation. For an explicit network model, parameters used for a power flow analysis is required. However, to reduce complexity during the optimization step, this model is simplified and the input parameters are also simplified accordingly.

Typical data for transmission elements (lines, transformers) include the impedance, thermal capacity, a flag that states if the branch limits are enforced, monitored or ignored, and the device outage.

If the network is not explicitly modeled, the sensitivities of the branches with respect to the nodes can be used as input to represent these constraints in the formulation. In these cases, it is a regular approach to include those branches whose potential flows are either near or over their capacity limits.

7.3.1.6 Treatment of Non-Dispatchable Resources

Non-dispatchable resources in the system are increasingly growing. Therefore, such resources are gaining more and more attention recently in the well-established electricity markets. Examples of these resources are wind and solar energy generators. Non-dispatchable resources are defined as the resources that provide unscheduled energy to the market and hence it becomes necessary to set up proper market rules to consider them as "equitable and streamlined for curtailment." They are intermittent or variable resources and are not easily available on demand. The fact that they are non-dispatchable implies a possibility that a specific unit may not be available when needed in contrast to the case of dispatchable units. Market systems need to ensure that the growing penetration of such non-dispatchable resources can also contribute to the overall economic benefits of the system. Several markets allow the participation of these resources in the day-ahead and real-time markets. These resources can be scheduled based on a price curve which can include negative price bids, or can be self-schedule or price takers. Some markets require these resources to submit their energy forecasts to the markets. The accuracy of the forecast tool for these types of resources becomes a very important issue. Some markets already have or are planning to have sub-hourly changes. For example, the telemetry of non-dispatchable resources and short-term forecasts will be needed every 5 minutes or even less than that time interval. These additional tools and system changes can make it possible to incorporate these resources into the economic dispatch and allow them to set the market price.

7.3.2 Real-Time Energy Market

Generally, the real-time energy market or real-time market is a balancing market in which the market-clearing prices are calculated at fixed intervals of time using some kind of optimal economic dispatch algorithm (see previous chapters) based on the actual system condition. Actual system operating condition is typically provided by the *energy management system* (EMS) which includes a state estimator supported by data fed from every node in the power system. Unlike a day-ahead market, there are no pre-arranged trades or contract prices of power among participants in a real-time market. Power balance is enforced at all times and real-time prices have to be discovered at fixed intervals of time. Load customers or generators have literally no time to find the right price, but to follow the discovered prices that are determined by the market operator at fixed intervals. Therefore, the deviations in real-time from the contracted quantities of the day-ahead market or intraday market become the real-time market trades or contracts. The real-time energy markets in the United States are cleared every 3–5 minutes while the power balance is enforced at every second. This difference in time can be resolved either by shortening the time interval needed to clear the real-time market or by some kind of aggregation of cleared MW. Actual financial settlement is typically performed on hourly integrated market prices or LMP.

In terms of market model, the centralized pool approach is better in handling supply–demand balance and real-time market clearing than power exchange model. In US nodal markets, the balancing (real-time) price calculations are based on the concept of LMP. The balancing settlement is based on actual hourly (integrated) quantity deviations from day-ahead scheduled quantities and on real-time prices integrated over the hour. Wholesale load customers will pay the real-time LMPs for any demand that exceeds their day-ahead scheduled quantities and will receive revenue for demand deviations below their day-ahead scheduled quantities.

For example, assume that the wholesale load customer's day-ahead demand schedule is 100 MWh for a specific market interval (usually 1 hour). In real-time, assume that the actual consumption by that load customer is 105 MWh for the same interval.

Then, this load customer has to pay for additional 5 MWh consumed, based on the real-time LMPs. For the original day-ahead demand schedule of 100 MWh, the load customer will pay based on the day-ahead LMPs. For generators, they are paid the real-time LMPs for any generation that exceeds their day-ahead scheduled quantities and will pay at real-time LMPs for generation deviations below their day-ahead scheduled quantities. For instance, assume that the day-ahead generation schedule for a generator is 100 MWh for a specific market hour. In real-time, assume that this particular generator generates only 95 MWh. In this case, the generator has to pay back for 5 MWh deviation based on the real-time LMPs. Let us not forget that the generator already receives revenue for day-ahead schedule of 100 MWh based on the day-ahead LMPs.

Transmission customers also pay congestion charges based on the congestion price component of real-time LMPs for bilateral transaction quantity deviations from day-ahead schedules. Demand aggregators may self-schedule demand reductions for demand resources not dispatched in real-time by the system operator. All spot purchases and sales in the balancing market are settled at the real-time LMPs. Congestion that results from the real-time sales and purchases of energy is settled at the congestion price component of real-time LMP. Transmission losses that result from the real-time sales and purchases of energy are settled at the marginal loss component of real-time LMP.

In European markets, balancing market is an organized market carried out by the TSO where players with dispatchable generators and loads can make balancing bids. With balancing bids, participants also offer regulation services, that is, they offer to increase or decrease the power production (or consumption) for a given hour of operation.

In nodal markets, the LMPs from the day-ahead market form the forward prices while the LMPs from the real-time market form the spot prices. Theoretically, forward prices should closely track spot prices. Although the day-ahead market financially binds the market participants, it also allows them to secure some price certainty. This is because the price volatility observed in the real-time market is generally higher than the price volatility observed in the day-ahead market. Undoubtedly, the certainty of market prices is very important to any market participant. The more volatility the market prices exhibit, the more uncertainty the market participants will experience. That, in turn, will create greater risks for them. For that reason, the energy volume traded in the day-ahead market is much larger than the energy volume traded in the real-time market. Typically, in some markets, the energy volume traded in the day-ahead market can range from approximately 40 to 60% of total energy supplied or consumed for a typical day, while the energy volume traded in the real-time market can only range from 5 to 10% of total energy. Obviously, the energy volume traded in an electricity market has an inverse relationship with the volatility of energy prices. The rest of the energy volume is made up of bilateral transactions and self-schedules whose transaction prices are unknown to the market operator. There are higher levels of price certainty associated with bilateral transactions and self-schedules than those from both day-ahead and real-time markets. Therefore, some generators and load customers who are risk-averse prefer bilateral contracts and self-schedules.

7.3.2.1 Real-Time Market Operation

Real-time market operation is intricately tied with real-time system operation. In real-time system operation, the instantaneous demand is instantly supplied by real-time energy production. If demand falls, supply is reduced almost instantaneously. Ancillary services are also offered as needed and demand is curtailed as necessary in extreme conditions (aka load shedding). Real-time market operation relies on the following: the network model, the state estimation solution, the constraints, the short-term load forecast, the physical schedules and the updated supply offers, day-ahead schedules along with day-ahead bids, and newly submitted real-time bids. The real-time market-clearing engine calculates the generation dispatch for intervals that are generally between 5 and 10 minutes. It also establishes the net scheduled interchange values for the same period.

In some markets, the market opens the previous day and closes 1 hour before the trading hour, and results are published half an hour before the trading hour. The real-time unit commitment designates fast- and short-start units in 15 minute intervals and looks ahead 15 minutes. Short-term unit commitment selects short- and medium-start units every hour and looks ahead 3 hours beyond the trading hour every 15 minutes.

In real-time, the economic dispatch process dispatches imbalance energy, or the energy that deviates from the schedule, and energy from ancillary services. The economic dispatch algorithm runs automatically and issues dispatches in 5 minutes interval. Under certain contingency conditions, the market operator can dispatch for a single 10-minute interval.

The market subjects bid to market power mitigation tests and the hour-ahead scheduling process, which produces schedules for energy and ancillary services based on submitted bids. It provides final ancillary service awards and financially binding inter-tie schedules.

7.4 TWO-SETTLEMENT SYSTEM

Having these two markets—day-ahead and real-time markets—triggered another concept known as "two-settlement system." Two-settlement market system consists of two markets: day-ahead and real-time (balancing) markets. Two-settlement market provides market participants with the option to participate in a forward market for electric energy in the electricity market. Separate settlements are performed for each market. Each of the markets is explained in more detail in previous sections. Day-ahead market settlement is based on the scheduled hourly energy quantities and day-ahead hourly prices. Real-time market settlement is based on the actual hourly energy deviations from the day-ahead schedules, priced at real-time LMPs.

Let us use an example for the purchase of an airline ticket as an analogy to the two-settlement systems. If a traveler wants to travel to a destination, the traveler can purchase the airline ticket in advance. The advantage of purchasing the airline ticket in advance is that the traveler knows how much it costs to purchase that ticket. It is known as price certainty. The other option the traveler has is that he or she can wait until the day of the flight to buy the ticket. In this particular case, there is less certainty associated with the ticket price. Using this analogy, purchasing the air ticket in advance is similar to purchasing energy in the day-ahead market while purchasing the ticket on the day of the flight is similar to purchasing energy in the real-time market.

Benefits of two-settlement systems include (1) enhancing robust and competitive market in the respective control area and (2) providing additional price certainty to market participants by allowing them to commit and obtain commitments to energy prices and transmission congestion charges in advance of real-time dispatch (forward energy prices), submit price-sensitive demand bids, and submit increment offers, decrement bids, and up-to-congestion transactions.

Generally, in the two-settlement system, there are two possible conditions that are applicable to each load-serving entity and each generator:

1. The real-time demand of the load-serving entity exceeds its day-ahead scheduled quantity. In this case, the load-serving entity will pay the real-time LMP for any demand that exceeds their day-ahead scheduled quantity.

2. The real-time demand of the load-serving entity is below its day-ahead scheduled quantity. In this case, the load-serving entity will receive real-time LMP for any demand that is below their day-ahead scheduled quantity.

3. The real-time generation of the generator exceeds its day-ahead scheduled quantity. In this case, the generator is paid real-time LMP for any generation that exceeds its day-ahead scheduled quantity.

4. The real-time generation of the generator is below its day-ahead scheduled quantity. In this case, the generator will pay for any generation that is below their scheduled quantity.

Numerical examples are given for each condition. The price for both DA and RT is assumed to be total LMP of each respective market.[1] Examples 7.1 and 7.2 apply to a load-serving entity while Examples 7.3 and 7.4 apply to a generator.

In Example 7.1, note that the RT payment is based on the difference between DA demand and RT demand. In this case, the difference is 5 MW.

In Example 7.2, the RT payment becomes RT credit in this case because the actual demand consumption is less than the scheduled demand consumption from the DA market. The payment for the load-serving entity is less than that in the previous example.

In Example 7.3, the extra 5 MW generation produced in RT market is settled by RT price, thus increasing the total revenue for the generator. In Example 7.4, the

---

[1] Unless otherwise stated, all prices given in dollars refer to US dollars

Example 7.1   Day-Ahead Demand is Less than Real-Time Demand

DA Demand (MW)	DA Price ($/MWh)	DA Payment ($)	RT Demand (MW)	RT Price ($/MWh)	RT Payment ($)	Total Charge ($)
100	20.0	2000	105	23.0	115	2115

Example 7.2   Day-Ahead Demand is Greater than Real-Time Demand

DA Demand (MW)	DA Price ($/MWh)	DA Payment ($)	RT Demand (MW)	RT Price ($/MWh)	RT Credit ($)	Total Charge ($)
100	20.0	2000	95	23.0	115	1885

Example 7.3   Day-Ahead Generation is Less than Real-Time Generation

DA Generation (MW)	DA Price ($/MWh)	DA Revenue ($)	RT Generation (MW)	RT Price ($/MWh)	RT Credit ($)	Total Revenue ($)
200	20.0	4000	205	22.0	110	4110

Example 7.4   Day-Ahead Generation is Greater than Real-Time Generation

DA Generation (MW)	DA Price ($/MWh)	DA Revenue ($)	RT Generation (MW)	RT Price ($/MWh)	RT Payment ($)	Total Revenue ($)
200	20.0	4000	190	22.0	220	3780

generator earned less because its RT generation is less than the scheduled generation from DA market.

7.5 MARKET OPERATION TIMELINE

In this section, we will illustrate examples related to the timeline process for different markets. We will describe the computational tools that are used for each process for both real-time and day-ahead markets. The main objective of these descriptions is to allow the readers to have a better understanding of the accuracy and computational performance involved in this mechanism.

For the day-ahead markets in Europe, participants submit their bids or offers to the power exchange markets until noon of the day before. Then, the calculation and checking process starts. This process guarantees the energy balance and the system security. First, the restriction calculations are performed until 2 p.m., and

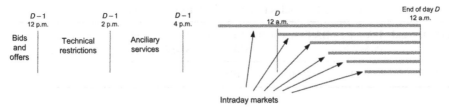

Figure 7.1 Time Sequence for Markets in Europe.

then, the ancillary service market starts and finishes at 4 p.m. With this information, the adjustment market named intraday market begins on the previous day and finishes on the market day after the delivery. It is performed on a 4-hour basis, Figure 7.1 gives an example of the sequence.

In some European markets for example, the operation of both market and power system is coordinated by two seperate institutions: a *market operator* (OM) responsible for the financial part and a *network operator* or *system operator* (OS) which verifies the technical feasibility of the system operation based on the market outcome. Market products include physical bilateral contracts for both medium and long term, daily market, and intraday market. Also complementary services are traded in a different market to comply with necessary conditions for quality, reliability, and safety set by the regulations.

Sale and purchase bids are included in a matching process for the day-ahead market. Market participants submit bids for each unit of production they own and for each market period. It comprises simple offers and complex ones. Simple bids have a price and an amount of energy and can be defined for each period within the horizon of production up to 25 energy blocks and with a different price for each block, in increasing manner. The offers do not include additional conditions to be taken into account in the matching of supply and demand.

Complex bids are the bids that meet the conditions of simple offers and also incorporate one or more of the following conditions: indivisibility, minimum income, scheduled off status, or variation of production capacity or load gradient. From the algorithm point of view, the market-clearing steps are as follows:

1. Determination of demand and supply curves per period
2. Search for the first valid solution
3. Simple conditional matching: Gradients (up, down, forward, backward)
4. Minimum income condition
5. Successive improvement of the first valid solution

This process is solved iteratively and the final results are expected to be obtained in less than 30 minutes.

Using Midcontinent ISO market in the US as an example, the input data is received until 10:30 a.m. a day prior to the actual market operation, and the results are posted

Figure 7.2   Time Sequence for Markets in the United States.

by 1:30 p.m., during which an SCUC, SCED, and a feasibility test are performed. Subsequently, a reliability assessment commitment is done, and the commitment notification is posted at 6 p.m. A simple schematic of this process is illustrated in Figure 7.2. These times are given as an illustrative example since different markets set them differently, especially, the due time for offers and the time when the day-ahead market prices are posted. Table 7.1 shows these time periods for some of the electricity markets in the United States. Note that these market time periods can change in the future for various reasons.

In the real-time market, the generation dispatch is performed every 5 minutes, after which the locational marginal prices are calculated and published every 5 minutes. As a general description for the real-time market, the initial base schedules and energy bids are due 75 minutes before the operating hour. One hour before, the market operator checks the energy balance, the network congestion and if there is enough ramp capability. If there are changes needed, the new schedules are submitted in the next 5 minutes. The market operator has 10 minutes to perform a new set of calculations. Then, it executes the calculations for the first 15 minutes of the hour. For these calculations, the market uses the load forecast, the bids, and the predefined outages. Twenty minutes before the corresponding time, the generator schedules and the corresponding prices for the first 15 minutes of the hour are posted. Then, a dispatch is calculated continuously every 5 minutes for the next 5-minute period, and the optimal power flow is computed for the next 15-minute markets and normally for additional periods up to the next 4 or 5 hours. Figure 7.3 illustrates this process.

7.6 TOOLS FOR MARKET OPERATION

System operators use different analytical tools to operate the different markets, and take decisions related to the commitment, scheduling, price, and system security. As

Table 7.1   Day-Ahead Market Timelines in the United States.

Market	Bids Deadline	Price Notification
CAISO	10 a.m. (PPT)	1 p.m. (PPT)
PJM	10:30 a.m. (EPT)	13:30 p.m. (EPT)
NYISO	5 a.m. (EPT)	11 a.m. (EPT)
MISO	10 a.m. (EPT)	13:30 p.m. (EPT)

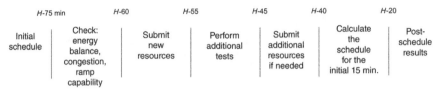

Figure 7.3 Time Sequence for Real-Time Market.

an illustration, security-constrained economic dispatch, reliability unit commitment, and the network security analysis are briefly presented.

7.6.1 Security-Constrained Economic Dispatch

Based on the general formulation of the economic dispatch, different markets consider different objectives. For example, the economic dispatch applied to real-time markets can perform different functions: (1) calculate the MW imbalance requirement for the next period, including the set point for each dispatchable generator, (2) determine the locational market prices for the next interval, and (3) compute the ancillary service capability of each unit and the ancillary service schedules. The types of objectives as well as constraints change based on the particular market design, and time horizon of the market. Then, the tools alter the name of the SCUC application according to the problem that needs to be addressed. Some of these commitment processes include reliability unit commitment, residual unit commitment, short-term unit commitment, and real-time unit commitment.

7.6.2 Reliability Unit Commitment

Reliability unit commitment (RUC) process is implemented to assess if the network system is operated in a reliable and secure manner. It also ensures that the system has sufficient level of generation resources and ancillary services to meet these requirements. Furthermore, it makes sure that the commitment selects the units in the correct locations to satisfy the forecasted demand. The RUC objective function is to minimize the total operation cost subject to the system and resource security constraints. The RUC settlement uses a payment mechanism to ensure that all generating resources committed by RUC are adequately compensated for their operation costs. It is a daily or hourly process conducted to ensure sufficient generation capacity is committed to reliably serve the forecasted demand. In some markets, it is also used to monitor and ensure the transmission system security by performing a network security analysis.

The day-ahead market-clearing price is based on the voluntary energy offers and bids in addition to the forecasted load. The resources committed to this type of market may not be sufficient to meet the real-time energy and ancillary service capacity requirements. Hence, the RUC process is needed to procure enough resource capacity to meet forecasted load in addition to ancillary service capacity requirement. The RUC process works like a bridge filling the capacity gap between the time when the calculations are done in the day-ahead market and the time before the real-time market starts to ensure the reliable operation of the system. RUC then becomes an intermediate calculation between day-ahead and real-time markets. Initially, the units that are needed for reliability are added after the posting of the day-ahead market results and not as part of the unit commitment. Markets start including the reliability units directly into the unit commitment for the day-ahead market. This inclusion, in general, reduces the difference between the day-ahead and the real-time market results. Some markets execute a multipass SCUC in order to satisfy reliability requirements.

7.6.3 Network Security Analysis

Real-time applications require fast and reliable computation methods due to the high number of possible outages in a modern power system. However, there is a well-known conflict between the accuracy of the method applied and the calculation speed. Therefore, several electricity markets use DC load flow-based approximate methods to quickly identify conceivable contingencies, and AC power flows only for critical contingencies.

7.6.4 Consideration of Transmission Constraints

When the system is congested, market operators need to take corrective actions such as dispatching additional generating units, since the system needs to operate in a secure manner. This correction generally affects power transaction curtailments and the increase in market prices. To consider and model the transmission network constraints into the problem, different markets use different methods depending on the tools they have at their disposal and the criteria on the accuracy or speed they require. Generally, these methods fall into three categories as follows:

- **Uninodal Method**: Transmission constraints are not considered directly. Once the dispatch is calculated, the flows are verified and corrected with the assumption that the transmission system may not affect the market. This type of approach can be suitable for network systems with a low level of congestion.
- **Explicit Network Model**: The marginal price calculation and dispatch algorithm include the power flow model and contingency analysis, considering the

network congestion as part of the algorithm. This methodology is typically adopted if the system experiences a high level of congestion.

- **Area Model**: The market is divided into zones or areas, and the model within each area is considered as uninodal. Only the interchange between the areas is monitored for potential congestion. It is an alternative between the explicit network model and total uninodal model.

7.7 FINANCIAL TRANSMISSION RIGHTS MARKET

The market equilibrium price or the market price is the foundation of any market. It is the price that provides incentives and guidance to motivate the market participants to behave in such a way that helps achieve the market outcome that is desirable to both the system operator and market participants. However, market price is not static. Market prices are variable, fluctuating, and volatile. The prices in the electricity markets are no exception to this phenomenon. As both day-ahead and real-time markets produce nodal prices (or zonal prices) at every possible node (or zone) in the market system, these prices can vary, sometimes significantly. There are several reasons for this price variation. The price variation in an electricity market is, in part, driven by the transmission congestion that is occurring somewhere in the system or, in part, driven by the variation in the marginal generation costs at different times.

Due to this price variation, the transmission congestion charges for transacting energy in the market are uncertain. As a consequence, the transmission customer may incur fluctuating charges over time. In addition, to improve the network utilization, mainly for highly congested systems that represent a problem for the market, an additional market related to the use of the transmission system is included. To hedge against those fluctuating charges, the customer can acquire suitable *financial transmission rights* (FTRs) which are available in nodal markets. FTR is equivalent to the right to transfer, inject, and withdraw power from the network. It is a type of financial risk management instrument. The acquired FTR can provide the customer with FTR congestion payments from the market that can be used to offset the transmission charges which can arise from the market price fluctuation.

FTRs, also known as *congestion revenue rights* (CRR), *transmission congestion contracts* (TCC), and *transmission congestion rights* (TCR) in some markets, are financial instruments that allow the market participants to hedge against the cost and uncertainty that may arise from the congestion through the use of transmission in the day-ahead market. An FTR comes in two types: *obligation* and *option*. An FTR obligation contract entitles the contract holder either to be charged or to receive compensation based on that congestion over the FTR path in the day-ahead market. If the congestion between specific points on the grid in the day-ahead electricity market has the same direction as the FTR path that a market participant holds, the contract holder will receive revenues. If the congestion is in opposite direction, the contract

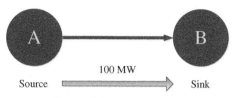

Figure 7.4   FTR Example.

holder will be charged. For FTR option contract, the contract holder is not liable for charges if the congestion is in opposite direction. Because there is no charge for FTR option holder, it is generally more expensive to acquire FTR options than FTR obligations.

Both FTR obligations and options are obtained for source–sink pair or point-to-point (receipt point-to-delivery point) pair. FTRs are defined by their MW quantity, source–sink pair and set for a specific time (monthly, seasonal, or yearly). FTRs can source or sink at the following types of points: generation node, hub, load zone, and interface point. FTR term has start and end dates. FTR periods are classified into peak/off-peak periods. An example of FTR is given in the Figure 7.4. This FTR has a quantity of 100 MW, from source A to sink B. This FTR is valid for only certain hours, say all peak hours of the month of July. FTRs are settled based on the difference between the marginal congestion components of the day-ahead LMPs at the source and sink points.

Possession of FTRs allows the FTR owners to hedge against congestion charges through congestion credits and are sold and bought in centralized auctions conducted by the market operator. The auctions can be based on estimated peak load, firm point-to-point transmission services, or through monthly secondary auctions.

The objective of the FTR auction is to maximize the revenues considering the network transfer capability and including the flows produced by the bids at different points of injection and withdrawal from the system. In general, it is calculated using a modified version of a security-constrained optimal power flow. The calculations may include auctions, initial FTR allocations, auction revenue rights (ARR), expansion FTRs, and revenue shortfall analyses.

To ensure the system security, the SCOPF optimizes the transmission rights not only by considering the base network but by modeling the contingency states, typically the "$N - 1$" security mode. All the implementation of FTR auctions may differ in the objective functions and constraints that vary according to the particular market applications. Some of the applications use "DC" network models. In terms of computing cost, the performance of the SCOPF application to solve a FTR problem is worse than that of the conventional SCOPF calculation. This difficulty is caused by the utilization of FTR options and the constraint models are non-sparse. Therefore, computation costs are considerable and increase exponentially with of the size of the power system.

The value of FTR obligation can be written as

$$V_{\text{obligation}} = \left(\text{MCC}_{\text{sink}} - \text{MCC}_{\text{source}}\right) \times \text{FTR}, \qquad (7.1)$$

where $V_{\text{obligation}}$ is the value of FTR obligation, MCC_{sink} is the marginal congestion component of the sink node LMP, $\text{MCC}_{\text{source}}$ is the marginal congestion component of the source node LMP, and FTR stands for MW amount of FTR that a market participant holds.

Let V_{option} represents the value of FTR option. Then, V_{option} can be expressed as

$$V_{\text{option}} = \begin{cases} (\text{MCC}_{\text{sink}} - \text{MCC}_{\text{source}}) \times \text{FTR} & \text{if } (\text{MCC}_{\text{source}} < \text{MCC}_{\text{sink}}) \\ 0 & \text{if } (\text{MCC}_{\text{source}} > \text{MCC}_{\text{sink}}) \end{cases}$$

Note that the value of FTR option (V_{option}) will be zero if ($\text{MCC}_{\text{source}} > \text{MCC}_{\text{sink}}$). In this case, the direction of the congestion is opposite to that of FTR path.

FTRs are financial instruments, not physical rights. FTRs do not give owners the right to schedule the physical use of the system. They are independent of fees associated with transmission service reservations. FTRs are a hedging mechanism that allows the FTR holder to manage the risk of congestion charges that may arise from the use of the transmission system in the day-ahead market. They do not protect the holder from congestion charges related to scheduling power in real-time or deviation from the day-ahead schedule. They also do not hedge against transmission loss charges. Credits are paid to the FTR owner provided that there are congestion rents collected to fund all target congestion credits for FTRs.

The high level FTR auction process is shown in Figure 7.5. The auction is the central mechanism of an FTR market. Each auction deals with FTRs that are valid over a specific time period, such as month, season or year as well as peak hours, off-peak hours, or both. All FTR auctions in US markets are sealed-bid type. The objective of FTR auction is to maximize the auction revenue received from buyers and sellers subject to the simultaneous feasibility with $N - 1$ security MW limits on all monitored elements. FTR holders can offer their FTRs to sell and FTR buyers bid

Figure 7.5 FTR Auction Process.

to buy FTRs in the FTR market. After the auction clearing, winning FTR offers are accepted and winning FTR bids are awarded. Market-clearing prices of all FTRs are produced as part of the auction process.

7.8 CAPACITY MARKET

Generators have both fixed costs and variable costs. If a generating unit is selected to run in a day-ahead energy market, the price (hence revenue) that the generator will receive is supposed to cover its variable costs. In fact, the generator needs to earn a level of revenues that will cover not only its variable costs but also its fixed costs to maintain its business viability. The additional revenue that a generator receives in an energy market above and beyond its variable costs is known as *inframarginal profit*. Generators have to rely on their inframarginal profits to cover their fixed costs. For a marginal unit, assuming it is bidding at its marginal cost, the unit will earn zero economic profit because the price it will receive equals its variable costs. In other words, there is no extra profit for the generator to cover its fixed costs. In a traditional rate-based regime, the capital costs of generating resources were recovered through retail rates when the power plants were owned by the state-regulated utilities. With the introduction of competition in the generation sector, generators are exposed to short-term efficient prices from the market environment.

Energy markets, including both day-ahead and real-time, operate based on the premise of achieving short-run economic efficiency. In a short-run economically efficient market, the generators are dispatched based on the merit-order to meet the required system demand. The merit-order means that the cheapest generators will be dispatched first before more expensive generators are dispatched. Most of the energy markets were able to achieve that goal. As high prices in energy markets can result from either a scarcity condition or exercise of market power by some generators, market operators implement *offer cap* or *price cap* in energy markets to protect consumers from unusually high price. Offer cap means a cap on the maximum offer made by each generator, while the price cap essentially provides a ceiling for the final market price cleared from the market. Price cap is not the cap on the generator offer, hence is different from offer cap. Again, in nodal markets which consider transmission constraints, the actual market prices can be higher than the maximum allowable offer cap. For example, if the offer cap is $1000/MWh, the final market price at a node can be higher than $1000/MWh.

However, because of such offer cap rules in certain markets, market prices can only go up as high as possible allowed by the offer cap. In actual nodal markets, market prices can go higher than the level of offer cap due to the nature of the mathematical formulation used in the market clearing. That means that there is an upper bound in terms of how much revenue that each generator can earn. At times, due to that price upper bound in combination with the limited operating hours that some units actually ran, the inframarginal profits may not be sufficient to cover the fixed costs for some generators. For those generators, relying just on the revenues from the energy markets only appears to be insufficient for the long-term viability of their business. This problem is known as *missing money* problem.

In some energy-only markets, one way to resolve this issue is to allow the offer cap to go as high as possible. For example, ERCOT, an energy-only market, received approval from the Texas state regulator to raise their system-wide offer cap to $9000/MWh in 2015 from $4500/MWh in 2012. The raise of offer cap in ERCOT was gradual in the sense that the offer cap in 2013 was $5000/MWh and that in 2014 was $7000/MWh. Similarly, *Australian National Electricity Market* (NEM) has a rule that sets a maximum spot price, also known as a market price cap, of AUS $12,500/MWh. This is the maximum price at which generators can bid into the energy market and is the price automatically triggered when *Australian market operator* (AEMO) directs network service providers to interrupt customer supply in order to keep supply and demand in the system in balance.

However, in some other markets, instead of raising the offer cap or price cap to a very high level, capacity markets were created to provide additional revenues for generators which can be available and can be required to operate in certain peak conditions. Electricity markets in PJM and ISO New England fall into that category. For example, PJM has a rule that sets an offer cap of $1000/MWh for every participating generator for many years. However, this $1000/MWh offer cap was recently raised to $2000/MWh under certain circumstances, such as when a participating generator has justifiable cost above $1000/MWh. Raising the offer cap in energy market as high as possible is not always an acceptable option in public opinion. Thus, the capacity market is designed to ensure that the market system has sufficient generating resources to reliably meet the peak demand by providing a payment towards the fixed capital costs of these resources. The creation and development of a capacity market is aimed towards to resolving that missing money problem. The capacity market is explicitly designed to provide resource adequacy so as to ensure the reliability of the system.

7.9 ANCILLARY SERVICE MARKETS

The operation of a power system requires the provision of some extra services in addition to the generation and transmission of energy to meet the load. These additional services are collectively known as *ancillary services* (A/S). In addition to the requirement to produce real power energy to meet consumers' demand, A/S is necessary to support the transmission of electric energy between selling and purchasing entities so as to maintain the reliable operation of the system. The main drivers for this A/S requirements are instantaneous change in load, planned and unplanned outages of generators, unforeseen system faults, and disturbances. A/S involves using generation capacities which are not currently used for providing energy to load customers. It can also involve load reductions in place of increased generation output. In the

United States, FERC Order 888 requires transmission providers to offer six ancillary services as follows:

1. Scheduling, system control, and dispatch service
2. Reactive supply and voltage control service
3. Regulation and frequency response service
4. Energy imbalance service
5. Spinning reserve
6. Supplemental reserve

There are two approaches to procure ancillary services:

1. **Cost-Based**: In the cost-based method, the services are offered at the pre-determined regulated costs.
2. **Market-based**: In the market-based method, services are provided at prevailing market prices determined by the market mechanisms which are approved by state or federal regulators.

The A/S services that are necessary for the reliability and integrity of the system are as follows:

1. Regulation service
2. Synchronous reserve
3. Non-synchronous reserve
4. Black start service
5. Reactive power service
6. Voltage support and regulation

While market mechanisms currently exist for regulation service, synchronous and non-synchronous reserves in many markets, it is challenging to establish markets for black-start, reactive power, and voltage regulation services. Those ancillary services that do not have market mechanisms are still provided by the cost-based method.

7.9.1 Regulation Service Market

In a power system, the load changes instantaneously. A flip of a switch by customers would change the load in that instant. Sometimes, the magnitude of change in load for a control area is very small (less than 1 MW). This instantaneous change in load has to be met by instantaneous change in generation even if the change is a very small amount. The imbalance of load and generation at any moment will cause the system frequency to deviate from its nominal value (50 or 60 Hz) and can cause under-frequency or over-frequency event which can lead to load shed or generator shutdown. Regulation refers to the control action that is performed to correct for load changes that may cause the power system to operate below or above its nominal frequency. By having a near balance of load and generation, the system frequency will be within acceptable limits, thus making the system balanced and stable. System stability is important for the integrity and robustness of the system, so as to avoid blackout. The change required of generators (and load) for balancing generation and load is called *regulation service*. As a general rule, a generating resource must meet certain criteria, as outlined below, to provide a regulation service:

1. The generating resource must be able to provide 0.1 MW of regulation energy instantaneously, when called by the ISO or the system operator.
2. The generating resource must have a governor capable of AGC control.
3. The generating resource must be able to receive an AGC signal from ISO. Resource MW output must be telemetered to the ISO control center in an acceptable manner.
4. Resources must demonstrate that they meet minimum regulation performance standards by passing the regulation performance tests.
5. Resources must give priority to the regulation signal (by ISO) by maintaining regulation capability at all times they can provide energy to the system.

Assume the forecasted off-peak load for an operating day is 68 MW, while forecasted on-peak load for the same operating day is 195 MW (arbitrarily chosen). Assume that the reliability regulation requires that regulation requirement for both off-peak hours (12 a.m. to 5 a.m.) and on-peak hours (5 a.m. to 12 a.m.) should be 0.70% of forecasted off-peak and on-peak, respectively. So, for the given example, the off-peak regulation requirement would be 0.476 MW, while the on-peak regulation requirement would be 1.365 MW.

The following types of resources can provide regulation: steam turbine, hydroelectric, combustion turbines, and combined cycle resources. Batteries, flywheels, plug-in hybrid electric vehicles, and various kinds of demand responses can also provide regulation service. Generating resources that plan to provide regulation service are required to pass some performance tests to prove that they are capable of doing so.

Using a simple numerical example, let us illustrate the compensation for a generator for providing a regulation service. Assume that a generating resource provides 1 MW of regulation service for the entire operating day (24 hours equivalent). That resource will be compensated for the variable cost of providing 1 MW of energy. The variable cost function of a generating resource includes fuel cost, emission cost, and variable O&M. The fuel cost is defined by the product of fuel price ($/MMBtu) and heat rate (MMBtu) of the unit. A simple numerical example is shown below.

Assume the following for this generator: (1) heat rate of the unit = 10,000 Btu/kWh, (2) fuel price (natural gas) = \$5/MMBtu, (3) emission cost = \$3/MWh, and (4) O&M cost = \$3/MWh. Hence, the fuel cost becomes \$50/MWh (= 5 \$/1,000,000 Btu × 10,000 Btu/kWh × 1000 kW/MW). Hence, the total variable cost of that unit becomes \$56/MWh (50 + 3 + 3). Hence, the cost of providing 1 MW of regulation energy for the entire day from that unit would be \$56.

The next step is to allocate that regulation service cost to the respective market participants. First, assume that all load customers will pay for the regulation service cost. In this case, that regulation cost of \$56 will be allocated among load customers based on their real-time load consumption. If both load consumers and generating units are equally considered, half of \$56 (= \$28) will be allocated to all load consumers and the other half will be allocated to all generators.

In PJM system, all LSE have hourly regulation obligation. The pro-rata share of hourly PJM regulation is based on LSE total real-time hourly load. Generally, LSEs can satisfy their regulation obligations by self-scheduling their own resources, entering bilateral transactions with other participants, or procure them from the regulation service market.

7.9.2   Primary Reserve Markets

Generally, a power system has to maintain a certain quantity of energy reserves on the system that will be available in 10 minutes to counteract a sudden loss of a generator or other contingencies. Traditionally, this quantity is roughly 150% of the largest contingency on the system. The largest contingency means the MW amount of the largest generator in the system. In the Eastern Interconnect of the United States, the largest contingency is approximately 1300 MW which is the capacity of the largest nuclear plant. This total 10-minute energy reserve is called *primary reserve*. Primary reserve has two components: *synchronous reserve* and *non-synchronous reserve*.

7.9.2.1 Synchronous Reserve Market   Synchronous reserve, also known as *synchronized reserve* or *spinning reserve*, is a reserve capability in the system that can be fully converted into the generated energy or load that can be removed from the system within 10 minutes and must be provided by equipment, such as generators or demand response which are electrically connected to the system.

Any power system needs a synchronous reserve as a way to counteract and control the impact of any disturbance anywhere in the system. Examples of the disturbances include a sudden loss of generator, loss of load, and loss of network components. If there is a sudden loss of generator, the supply becomes deficient, causing system frequency to dip. In a few seconds, the generators which are online (hence *synchronous*) will automatically react by producing more power to fill the void. The MW amount of synchronous reserve needed in a system and which generating resources will provide how much of synchronous reserve depend on the characteristics and operational requirements of the system. One of the mandatory requirements in the United States is that the system frequency must be restored back to the pre-contingency (contingency represents a disturbance) level within 10 minutes after the disturbance. So, it is important to maintain a sufficient level of synchronous reserve in the system, to restore the post-disturbance system frequency to a normally acceptable value.

A large power system can be broken down into a set of smaller zones with separate synchronous reserve requirements. The synchronous reserve sub-zone can also be defined depending on the system requirements. Generally, the minimum synchronous reserve requirement is defined as the largest contingency on the system. Largest contingency typically means the loss of the largest generating unit. The larger synchronous reserve may be required depending on the likelihood of other system contingencies. In order for a generating resource to provide the synchronous reserve, that resource must meet the following criteria at minimum:

1. The generating resource must be electrically connected to the power system.

2. The generating resources that are not electrically connected to the power system are not eligible to provide synchronous reserve.

3. The generating resource must be online following economic dispatch and able to ramp up from their current output in response to a synchronous reserve event within 10 minutes.

4. The eligible resource will be compensated for the amount of MW response they provide.

Synchronous reserve requirements can be fulfilled through a market mechanism. The market mechanism is typically set up to procure synchronous reserve for next operating hour or next day. In some markets, reserve requirements for the system are cleared on a long-term basis. For example, ISO New England electricity market runs a market called *forward reserve market* (FRM). The goal of this forward reserve market is designed to attract investments in, and compensate for, the types of resources that provide the long-run, least-cost solution to satisfy off-line reserve requirements. The locational FRM compensates participants with resource capacity located within specific subareas for making the type of electric energy market offers that would make them likely to be unloaded and thus available to provide energy within 10 or 30 minutes. The ISO conducts two FRM auctions: one each for the summer reserve period (June through September) and winter reserve period (October through May), that acquire obligations to provide pre-specified quantities of each reserve product. Forward-reserve auction clearing prices are calculated for each reserve product in each reserve zone. When supply offers for forward reserve are not adequate to meet a requirement, the clearing price for that product is set to the price cap, which is $14.00/kW-month.

7.9.2.2 Non-Synchronous Reserve Market   Non-synchronous reserve, also known as *non-synchronized reserve* or *non-spinning reserve*, is a reserve

capability in the system that can be fully converted into the generated energy or load that can be removed from the system within 10 minutes and is provided by equipment not electrically synchronized to the system. The balance of primary reserve and synchronous reserve is generally made up by non-synchronous reserve. For example, if the primary reserve requirement of the system is 150% of the largest contingency in the system, the non-synchronous reserve requirement would be 50% of the same largest contingency while the synchronous reserve requirement is 100% of the same contingency.

The system operator will determine MW capability of each resource based on its operational characteristics and use this information to identify which units can provide non-synchronous reserve. The system operator will use some criteria to assign non-synchronous reserve MW commitment to certain resources in the system. Generally, the criteria will be based on the most economic set of resources to provide non-synchronous reserve. Examples of non-synchronous reserve resource generally include run-of-river, pumped hydro, industrial combustion turbines, jet engine/expander turbines, combined cycles, and diesel generators. Generally, the entire power system will have a single non-synchronous reserve zone. Additional smaller sub-zone for non-synchronous reserve can be defined depending on the characteristics and requirements of the system. Since synchronous reserve may be utilized to meet the primary reserve requirement, there is no explicit requirement for non-synchronous reserves. Non-synchronous reserve is eligible to be used to meet the difference between the primary and synchronous reserve requirements if it is economic.

In order for a generating resource to provide non-synchronous reserve, that resource must meet the following criteria at minimum:

1. The generating resource must be electrically connected to the power system.
2. The generating resource must meet the eligibility requirements to provide non-synchronous reserve. One eligibility requirement is to be able to increase in energy output within a continuous 10-minute period provided that the resource is not synchronized to the system at the initiation of the response.
3. Generation resources that have designated their entire output as emergency will not be considered eligible to provide non-synchronous reserves.
4. Generation resources that are not available to provide energy will not be considered eligible to provide non-synchronized reserves.

7.10 ADDITIONAL SERVICES

Reactive power, voltage regulation, and black start services are key components of reliable power system operation. However, market mechanisms for these services do not exist currently for various reasons.

7.10.1 Reactive Power Service and Voltage Control

By its nature, the voltages on the transmission facilities of a system must be maintained within acceptable limits. Voltages throughout the system can be controlled by adjusting the injection and withdrawal of reactive power at various locations. To maintain this system requirement, the system relies on the generating resources connected to the system to produce or absorb reactive power. In addition to the reactive power sources from generators, other reactive power devices, such as synchronous condensers, static var compensator (SVC), and settings of on-load tap-changing (OLTC) transformers, are also used. Thus, reactive supply and voltage control from generation sources must be provided for each transaction on the transmission provider's transmission facilities. The amount of reactive supply and voltage that must be provided with respect to the transmission customer transactions will be determined based on the reactive power support necessary to maintain transmission voltages within limits that are generally accepted in the region and consistently adhered to by the transmission providers.

Generation owners shall update reactive capability curves (D-curves) via some mechanisms to ensure that EMS security analysis results by the system operator and transmission owners are accurate. D-curve is also known as *continuous unit reactive capability curve* that provides the realistic usable reactive output that a generating unit is capable of delivering to the system operator and sustaining over the steady state operating range of the unit.

Typically, at the end of each month, the system operator will calculate the credits due each market participant for reactive services. Generators whose real energy output is altered at the request of the system operator for the purpose of maintaining reactive reliability within the system are credited hourly for lost opportunity costs if their output is reduced or suspended and credited in accordance with balancing operating reserve credit calculations if their output is increased. Generators operating as synchronous condensers for the purpose of maintaining reactive reliability at the request of the system can be credited in similar manner. The desired MWh used in this calculation is based on the hourly integrated real-time LMP at the generator bus and adjusted for any effective regulation or synchronized reserve assignments and is limited to the lesser of the unit economic maximum or the unit maximum facility output.

7.10.2 Black Start Service

Although black start is not defined as part of A/S, it is an essential service when the system faces a complete blackout. Black start capability is necessary to restore any power system following a blackout. Transmission system operator and owners of transmission and generation should designate a set of specific generators as black start units whose location and capabilities are required to re-energize the transmission system. These designated resources are generating units that are able to start without an outside electrical supply. The black start units can also include resources with a high operating factor which can demonstrate an ability to remain operative, at reduced levels, when automatically disconnected from the grid. The planning and maintenance of adequate black start capability for restoration of the balancing area of the power system following a blackout represents a benefit to all transmission customers. All transmission customers must therefore take this service from the system operator. Black start service can be provided by units that participate in system restoration. Such units may be eligible for compensation under the black start service. If a partial or system-wide blackout occurs, black start service generating units can assist in the restoration of the balancing area.

7.11 FURTHER DISCUSSIONS

In this chapter, the key components of well-functioning market designs are presented. However, there are other developing issues that are going to change some aspects of those design elements.

Basic market design in the United States is robust. The elements of this market design include day-ahead and real-time market models, nodal pricing, FTR trading, and capacity market. But there are some rooms for improvement. Those areas include capacity adequacy, uplift charges (from committing additional CT units), and gas-electric coordination. Risk is transferred from markets to reliability risks.

Some economists believe that it is not necessary to have a capacity market to ensure long-term reliability of the grid. They also believe that shifting to a capacity market would be a source of inefficiency and a barrier to competition that would likely increase the cost of electricity for consumers. For example, ERCOT is still debating about whether a capacity market, similar to those in other electricity markets in the United States, is needed to provide sufficient incentives for generation investment. Whether a capacity market is ultimately needed along with an energy-only market remains to be seen.

Theoretically, an energy-only market without a capacity market can still achieve the goal of system reliability. For example, if supply is going to be less than demand, the demand has to be reduced. This can be achieved by either activating DR programs, or simply shedding load. It is still a question of how much it costs for the system to shed load.

On the other hand, having a capacity market, along with well-functioning energy-only market, provides additional buffer or cushion to the challenging task of maintaining system reliability. Consumers have to pay additional cost in order to obtain that buffer. It is much like a premium payment for ensuring that the system is far away from the point of system collapse which can jeopardize the system reliability. It is a matter of choosing a particular market design.

Recent operating challenges in the US northeast, due to extreme cold weather, emphasized the need to coordinate electric supply and gas supply, called gas–electric coordination. The unavailability of gas-fired units due to gas supply interruption prompted PJM market operator to revise and revamp its capacity market construct, rewarding more to generating resources that will be available at system peak times and penalizing more to generating resources that will not be available.

Economists believe that markets are the best way to organize the provision of wholesale power. Overall market can be competitive, but local areas may not be, because generators may have market power. Market power mitigation is important so that participants behave competitively overall and the market outcome can be competitive.

CHAPTER END PROBLEMS

7.1 In every existing electricity market, bilateral contracts play a key role in matching the right buyer with the right seller at the right price. In real-time, the delivery of electricity takes place based on those bilateral contracts. Does supply of electricity to final consumers really need bilateral contracts? Why or why not?

7.2 Does an exchange only allow simple bids (one part bid of energy/price pair)?

7.3 Is bilateral market approach possible in real-time balancing market? why and why not?

7.4 Assume a market participant holds 100 MW of FTR from point X (source) to point Y (sink). The day-ahead market-clearing results showed that the total nodal price at node X is $27 with MCC $5 and MLC $2 and the total nodal price at node Y is $39 with MCC $15 and MLC $ 4. If the FTR is obligation type, what is the value of that FTR? What is the value of that FTR if the participant holds 200 MW instead of 100 MW? What would be the value if the FTR is option-type?

7.5 Generally, the market participants have to pay higher price to obtain FTR option than FTR obligation. Why?

7.6 As a forward market, only day-ahead markets exist today. Would it be possible to set up 3-day ahead or 7-day ahead markets? Why and why not? What are the pros and cons of setting up such forward markets?

FURTHER READING

1. Hogan WW. Contract networks for electric power transmission. *Journal of Regulatory Economics* 1992;4:211–242.

2. Chao HP, Huntington HG. *Designing Competitive Electricity Markets*. New York: Springer US; 1998.

3. Rosellón J, Kristiansen T. *Financial Transmission Rights: Analysis, Experiences and Prospects*. London: Springer-Verlag; 2013.

4. Stoft S. *Power System Economics: Designing Markets for Electricity*. John Wiley & Sons; 2002.

5. Einhorn MA. *From Regulation to Competition: New Frontiers in Electricity Markets.* Netherlands: Springer; 1994.

6. Rudnick H, Varela R, Hogan W. Evaluation of alternatives for power system coordination and pooling in a competitive environment. *IEEE Transactions on Power Systems* 1997;12(2):605–613.

7. Shahidehpour M, Yamin H, Li Z. *Market Operations in Electric Power Systems: Forecasting, Scheduling, and Risk Management.* John Wiley & Sons; 2002.

8. Buzoianu M, Brockwell A, Seppi DJ. A dynamic supply–demand model for electricity prices. Technical Report, Carnegie Mellon University, 2005. [Online]. http://repository.cmu.edu/statistics/134/

9. Brinckerhoff P. Electricity generation cost model. Department for Energy and Climate Change, UK, 2012. [Online]. https://www.gov.uk/government/uploads/system/uploads/attachment_data/file/65713/6883-electricity-generation-costs.pdf

10. Patrick RH, Wolak FA. Estimating the customer-level demand for electricity under real-time market prices. NBER Working Paper 8213, National Bureau of Economic Research, Cambridge MA, 2001. [Online]. https://www.nber.org/papers/w8213

11. Midcontinent Independent System Operator. [Online]. https://www.misoenergy.org/Training/Pages/Training.aspx

12. California ISO. [Online]. http://www.caiso.com/market/Pages/TransmissionOperations/Default.aspx

13. PJM. [Online]. http://www.pjm.com/markets-and-operations.aspx

14. OMI-Polo Espa nol S.A. (OMIE). [Online]. http://www.omie.es/inicio/mercados-y-productos

15. Alsaç O, Bright JM, Brignone S, Prais M, Silva C, Stott B, Vempati N. The rights to fight price volatility. *IEEE Power and Energy Magazine* 2004;2(2):47–57

Chapter 8

Pricing, Modeling, and Simulation of an Electricity Market

This chapter will cover three key aspects of an electricity market: pricing, modeling, and simulation. Pricing mechanism used in any electricity market is a fundamental and critical foundation on which the entire electricity market is built. The actual pricing currently used in the practical markets are more complex than what is described in this Chapter. In some advanced electricity markets, pricing includes pricing for price-responsive demands, reserve pricing, scarcity pricing, and pricing for other ancillary services. Modeling and simulation of an electricity market is also important as it can help us understand and answer many questions related to what-if scenarios and situations. Such questions include impact of merger and acquisitions, economic assessment of establishing or joining an electricity market regime, impact of structural change or market design change for an electricity market, and many others. Other policy-related questions can be answered as well. These scenarios can be modeled and simulated before their actual implementations are done in an actual market.

8.1 MARKET-CLEARING

Economic theory suggests that, in a free market, there will be a single price which brings supply and demand into balance, called equilibrium price. The buyers need to purchase the scarce resource that the sellers have and hence there is a considerable incentive to engage in an exchange. In its simplest form, the constant interaction of buyers and sellers enables a price to emerge over time. Market-clearing equilibrium

Electricity Markets: Theories and Applications, First Edition. Jeremy Lin and Fernando H. Magnago.

price is also called market-clearing price because at this price the exact quantity that producers take to market will be bought by consumers, and there will be nothing left over. This is efficient because there is neither an excess of supply and wasted output, nor a shortage—the market clears efficiently. This is a central feature of the pricing mechanism, and one of its significant benefits. For markets to work, an effective flow of information between buyers and sellers is essential.

In most electric power networks, demand and supply must be instantly balanced because storage of electricity is very expensive. The consequence is that transmission constraints and how they are managed in a market setting often have a large influence on market prices. In an electricity market, the market-clearing price is the price that balances the demand (the amount of electricity that buyers want to buy) with the supply (the amount that sellers want to sell). The market-clearing process for an electricity market can be described as a process in which an optimization problem is solved. For example, the real-time energy market is cleared through a *security-constrained economic dispatch* (SCED) algorithm. The objective of the SCED, which is an optimization problem, is to minimize the total production cost (or equivalently to maximize the total social welfare), subject to various system constraints to maintain system reliability. The results of SCED are desired resource output levels (dispatch schedules in MW) and market prices for either a zone or several nodes in the system (in $/MWh). Generally, there are two major pricing mechanisms employed in any electricity market: *zonal pricing* and *nodal pricing*.

8.1.1 Zonal Pricing

For most of the current electricity markets, a zonal pricing was the first pricing mechanism that was adopted to cope with network capacity problems. In a zonal pricing, the market price within a particular zone is uniform, without considering the possibility of transmission congestion inside the zone. If there are more than one zone involved, the inter-zonal transmission constraints are generally considered which can lead to different prices in different zones depending on whether these constraints are binding. The zone can be a large operating region or a state or a country. This zonal pricing design was thought to minimize the complexity of the market price settlement associated with nodal pricing. It is also politically more acceptable with a single uniform price within a state or a country. These are some of the main reasons why electricity markets in most European countries adopted the zonal pricing scheme. For example, when Norwegian power market was launched in 1994, a zonal pricing approach was adopted to manage congestion in the scheduled day-ahead power market. By the same token, electricity markets in the United States started with zonal pricing scheme when markets were launched in the beginning. However, US electricity markets later switched to the nodal pricing scheme which provides more efficient pricing. Some of the markets tried to gain market experience with zonal pricing for a number of years before they switched to the nodal pricing scheme. Electricity market in Texas (ERCOT) is one such market.

Generally, the market-clearing steps involved in zonal pricing are as follows:

1. Suppliers provide offers and demands submit bids (fixed bids and/or price-responsive bids). Based on these offers and bids, the market is cleared while ignoring any internal network constraints within a zone.

2. This market-clearing process produces a system marginal price of energy which represents the intersection of supply offers and demand bids (recall Chapter 3 on the determination of market-clearing price given supply and demand curves).

3. If the resulting power flows due to this market-clearing induce network capacity problems inside a zone, the nodes of the grid are partitioned into sub-zones.

4. If only two zones are considered, the zone with net supply is defined as the low-price zone, while the zone with net demand is defined as the high-price zone. This is because a more expensive supply resource in an import-constrained zone needs to be dispatched to meet that net demand.

While the zonal pricing scheme has the clear benefit of simplicity, there are several limitations and problems associated with that pricing scheme. One of the problematic issues with zonal pricing scheme was the difficulty associated with defining the zonal boundaries. First, any zonal definition would be quite arbitrary. The most-frequently used criteria in defining zonal boundaries is the transmission congestion which can cause price separation between perceived zones. Hence, redefining different zones due to dynamically varying congestion seems not practical. As is clear from the experience of subsequent nodal markets, the congestion can occur anywhere anytime in the system because the transmission grid is more complex than it can allow the zone to split into two or three separate zones. One logical way of defining zones would be based on the nodal price differences which comprise all relevant information on network related costs. Other criteria which use market concentration, liquidity, or transaction costs may not do better. In summary, it is difficult to define the number of zones and determine the zonal boundaries for an electricity market that adopts a zonal pricing scheme.

There are other drawbacks associated with zonal pricing. In the zonal pricing adopted earlier in US electricity markets, the system operator has to redispatch the generators whenever one or more transmission constraints within a zone were overloaded after zonal prices from real-time markets were announced. Redispatch of generators generally requires generation increase in import-constrained sub-zone and generation decrease in export-constrained sub-zone to relieve that transmission constraint. The compensation schemes used in this redispatch situation is typically outside the scope of zonal pricing scheme. Therefore, this post-market redispatch has no influence over how suppliers offer into the zonal market. Whenever a transmission congestion appears inside a zone, that congestion indirectly creates a situation in which the prices on the constrained and unconstrained sides of the zone have to diverge. That problem was explicitly resolved in the nodal pricing scheme by explicitly pricing out every possible node in the market system.

Several research works have also pointed out that nodal pricing is better than zonal pricing in preventing market power. The uniform price in a large zone would simply constrain the ability to detect market power that would be exercised by suppliers in that zone. The impact on social surplus and its allocation (surplus redistribution) based on nodal pricing versus zonal pricing would be significant. It is generally believed that a system with nodal prices provides correct investment signals due to price formation at specific nodes or locations within a zone or system. Additionally, the zonal pricing is no longer suitable for a market system with a growing number of variable generation resources, such as wind power, which are often located far from demand.

8.1.2 Nodal Pricing

The concept of spot pricing was first introduced by Prof. Fred Schweppe from MIT and his colleagues. They proposed a model in which the spot price is derived from a social welfare maximization problem subject to a number of constraints. The difference in spot prices between any two locations corresponds to a price of transmission costs between the two nodes. As electric networks are non-linear AC model, this is typically simplified by a linear approximation of the network using DC model-based load flow method. Later, DC load flow-based nodal pricing (called DCOPF) was developed which incorporates the technical externalities associated with transmission congestion and transmission losses. In this pricing scheme, tradable transmission capacity rights are adopted and trading rules that specify the transmission loss compensation necessary for power transfers are also introduced.

In essence, nodal pricing scheme is more complex and computationally intensive than zonal pricing. This is because nodal pricing uses the pricing scheme known as *locational marginal price* (LMP), in which electricity price is determined on marginal basis at the electrical bus level. In a large-scale electricity market which is based on a large-scale power system, there can be several thousands of electrical nodes. In the zonal pricing scheme, in contrast to nodal pricing, electricity price is determined still on the marginal basis, but at a larger zonal level. The granularity of price information available in a nodal pricing is generally not available in a zonal pricing method. Generally, zonal prices are formed as an aggregation of nodal prices at the bus level in some fashion, such as aggregation of load-weighted nodal prices.

Under nodal pricing, an LMP is generally composed of three components: *system marginal (energy) price* (SMP), *marginal congestion component* (MCC), and *marginal loss component* (MLC). In this case, the LMP at the chosen reference bus of the market network represents the SMP. Under standard decomposition approach, selection of reference bus is also important, because selection of two different reference buses could produce two different results for each component term in the standard LMP. The final LMPs are the same regardless of which node was chosen for the reference bus.

Nodal pricing or locational marginal pricing (LMP) method is currently used in electricity markets in Argentina, Chile, Ireland, New Zealand, Russia, Singapore, and all markets in the United States. This design explicitly acknowledges that location is an important aspect of electricity which should be reflected in its price, so all accepted offers are paid a local uniform-price associated with the network node at which each supply offer is located.

8.1.2.1 Mathematical Formulation

To set the stage for fundamental understanding of electricity pricing in the context of nodal electricity markets, we provide a basic, yet brief, mathematical formulation of nodal pricing here. The LMP at a location is defined as the marginal cost of supplying an additional increment of load at that location without violating any pre-determined system security limits. Using the solution from the *Lagrangian* formulation of an optimal power flow (OPF) problem, LMPs can be determined by using the *Lagrange* multipliers (shadow price) associated with equality and inequality constraints defined in the problem formulation.

The key input variables necessary for computing LMPs are generator offers, demand bids (fixed or price-responsive), generator-operating constraints (minimum and maximum limits of generation for each generator, ramp limits as applicable), transmission network parameters and topology, and line flow limits (minimum and maximum). Other constraints, such as voltage limits at each bus, or environmental constraints, are not explicitly included in the LMP computation. Environmental constraints, faced by each generator, are implicitly assumed to be modeled in the offer curves of generators.

The prevailing method of computing LMPs in current industry practice and mainstream research is to form the objective function using *Lagrange* formulation, based on the OPF-based market model. The general formulation of LMPs can be written as an OPF problem as follows:

$$\text{minimize} \quad F(p^g) = \sum_{i \in G} c_i(p_i^g) \tag{8.1}$$

subject to

$$-p^g + p^d + g(\theta) = 0 \quad \longleftrightarrow \quad \lambda^e \tag{8.2}$$

$$h(\theta) \leq \bar{s} \quad \longleftrightarrow \quad \mu \tag{8.3}$$

$$p_i^{g-} \leq p_i^g \leq p_i^{g+} \quad \longleftrightarrow \quad \eta_i^-, \eta_i^+ \quad \forall i \in G \tag{8.4}$$

$$p_i^g = 0 \quad \forall i \notin G, \tag{8.5}$$

where $\theta \in \mathfrak{R}^n$ is the vector of state variables (voltage angles of all nodes, except angle at specified node) and $g(\theta) : \mathfrak{R}^n \longmapsto \mathfrak{R}^{n+1}$ is the real power flow functions at all the nodes in $\mathcal{N}$ including losses due to nonzero branch resistance. The term $p^g(p^d) \in \mathfrak{R}^{n+1}$ is the vector of MW injections (or withdrawals) at each node, with $p_i^g \equiv 0$ for each node $i \notin G$. p_i^{g-} and p_i^{g+} denote the lower and upper limits of the generator's injection at node i. The function $h(\theta) : \mathfrak{R}^n \longmapsto \mathfrak{R}^{|B|}$ represents the real power branch flow as functions of voltage angles, and $\bar{s}$ represents the vector of limits for real power branch flows.

8.1.2.2 DCOPF-Based Formulation

All currently operating nodal markets in the United States are based on DCOPF formulation. The DC power flow model, which attempts to approximate the AC power system, is based on a number of assumptions, as described below:

1. Reactive power balance equations are ignored.
2. All voltage magnitudes are assumed to have the same 1.0 (p.u.) values at all buses.
3. Line power losses are ignored.
4. Dependence of reactance or series impedances of the LTC and phase shifting transformers on tap ratio values is ignored.

Based on these assumptions and using the similar notations used in the previous section, the OPF problem, based on DC power flow (called DCOPF) can be formulated as

$$\text{minimize} \quad F(p^g) = \sum_{i \in G} c_i(p_i^g) \tag{8.6}$$

subject to

$$-p^g + p^d + g(\theta) = 0 \quad \longleftrightarrow \quad \lambda^e \tag{8.7}$$

$$\mathbf{Hp} \le \mathbf{Z} \quad \longleftrightarrow \quad \mu_{dc} \tag{8.8}$$

$$p_i^{g-} \le p_i^g \le p_i^{g+} \quad \longleftrightarrow \quad \eta_i^-, \eta_i^+ \quad \forall i \in G \tag{8.9}$$

$$\mathbf{1}^T q = 0, \tag{8.10}$$

where $\mathbf{p} = [p_1 \quad p_2 \quad \cdots \quad p_n]^T$ is the nodal active power injection vector; $\mathbf{H}$ is an $m \times n$ matrix consisting of the submatrix of *power transfer distribution factors* (PTDFs), also known as *generation shift factors* (GSF), corresponding to the transmission constraints and the submatrix representing the capacity constraints for non-slack buses; $\mathbf{Z}$ consists of the transmission capacity limits and the generation capacity limits for non-slack buses.

As mentioned earlier, the LMP is generally composed of three components: the system marginal price (SMP), the marginal congestion component (MCC), and the marginal loss component (MLC). The *Lagrange multiplier* for power balance constraint (λ^e) in Eqs. (8.2) and (8.7) represents the system marginal price (SMP) for meeting the demand, while the *Lagrange multipliers* for inequality constraints (μ), such as line flow limits, in Eqs. (8.3) and (8.8) represent part of marginal congestion component (MCC) in the LMP formulation. Thus, the LMP at any node i (λ_i) can be mathematically defined as

$$\lambda_i = \lambda^e + \lambda^c + \lambda^l \tag{8.11}$$

In the above formulation, the λ^l represents marginal loss term. The marginal congestion component of the LMP (λ^c) is the product of the *Lagrange multipliers* for inequality constraint j, represented by shadow price μ_j, and shift factors SF_{jk} for each non-marginal bus k:

$$\lambda^c = \sum_{j \in N} \mu_j \times \text{SF}_{jk} \tag{8.12}$$

Shift factors can be mathematically expressed as

$$\text{SF}_{jk} = \frac{\Delta F_j}{\Delta P_k}, \tag{8.13}$$

where SF_{jk} are shift factors for the line j with respect to the bus k, ΔF_j is the change in MW flow on the line j when the change in real power occurs at bus k and ΔP_k are changes in MW at bus k. Shift factors represent linear sensitivity factors representing the change in MW power flow on a particular line, with respect to the change in unit MW at a particular bus. The bus can be a load or a generator bus, and the change in unit MW is usually an increment of 1 MW at that bus. It is assumed that the change in unit MW is compensated by the opposite change in MW at the reference bus in the system. Shift factors can be derived from either DC or AC power flow.

8.1.2.3 Marginal Loss Component
The nodal pricing is based on the choice of a reference bus or load-weighted distributed slack bus. By definition, the locational price of energy at that reference bus represents SMP and is used in calculating nodal prices for all other nodes across the market system. The loss and congestion components at the reference bus, by definition, are zero. At other nodes, the state of transmission grid determines the transmission losses and transmission constraints determine the congestion values.

Losses are due to the electrical resistance (or impedance) in transmission lines and transformers. Losses incurred are low for generation nodes which are closer to load centers compared to generation nodes in regions which are more distant from the loads. If there were no constraints in a market system, then, the congestion component would equal zero and price differentials among pricing nodes would be due entirely to the cost of losses. Generation that is close to load centers is most valuable at these times as the losses incurred in getting the energy to the customers is very low and, in some instances, the incremental energy will reduce the total system losses. The generation that are most distant from the loads is worth the least as the losses incurred getting the energy to the loads would be large.

Losses occur in the transmission system as energy flows from generation sources to the loads. These losses appear as additional electrical load, requiring the generators to produce additional power to compensate for the losses. The amount of losses that occur on specific transmission lines or areas of the transmission network at any given time is dependent on the network topology and the specific generation sources being used to meet the load at that time. Two important factors are defined as follows:

A: Penalty Factor

The penalty factor for generator i (PF_i) is defined as the increase required in generator output at bus i (ΔG_i) to supply an increase in load at the reference bus (ΔD) with all other loads held constant:

$$PF_i = \frac{\Delta G_i}{\Delta D} \tag{8.14}$$

This penalty factor can also be defined as follows, based on the energy balance relationship:

$$PF_i = \frac{1}{1 - \frac{\Delta \text{Loss}}{\Delta G_i}} \tag{8.15}$$

Generator energy bid prices are multiplied by penalty factors to account for incremental transmission losses in the dispatch process.

B: Delivery Factor

The delivery factor of a generator i (DF_i) is defined as below:

$$DF_i = \frac{\Delta D}{\Delta G_i} \tag{8.16}$$

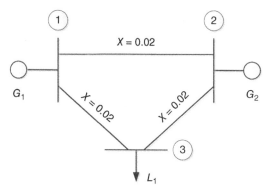

Figure 8.1   Three-Bus Network System.

which is the reciprocal to the penalty factor:

$$DF_i = \frac{1}{PF_i} \qquad (8.17)$$

Delivery factors are used to calculate the marginal loss components of the LMPs. Since $\lambda^l = [DF_i - 1]\lambda^e$, Eq. (8.11) for node i becomes

$$\lambda_i = \lambda^e + \lambda^c + [DF_i - 1]\lambda^e \qquad (8.18)$$

8.1.2.4 Three-Bus Example   To illustrate the fundamental calculation of LMPs for a power system, we use the following three-bus network system as an example, shown in Figure 8.1, with the corresponding network data. There are two generators in the system, with one generator located at bus # 1 (unit # 1) and another generator at bus # 2 (unit # 2), with demand at bus # 3. We assume that the generator at bus # 1 (unit # 1) is a cheap coal-fired unit with the marginal cost of $20/MW with a capacity of 100 MW. The generator at bus # 2 (unit # 2) is an expensive gas-fired unit with the marginal cost of $40/MW with a capacity of 100 MW. The initial demand value is assumed to be 100 MW (bus # 3). Bus # 1 is assumed to be the reference bus.

A: Unconstrained Network Case

Given the network and generator offer information, we can easily determine the nodal prices for the system if there is no network congestion in the system and if marginal losses are ignored. The nodal prices for this unconstrained network for different levels of demand at bus # 3 are shown in the Table 8.1 below. LMPs are undefined for a load level of 200 MW and above, because there is no additional MW available to meet that one incremental MW at the load bus.

Table 8.1 Nodal Prices ($/MWh) for Unconstrained Network for Different Load Levels

Load Level (MW)	Bus # 1	Bus # 2	Bus # 3
0–99	20	20	20
100–199	40	40	40

Assume again that the load level at bus # 3 is 100 MW. Based on the aggregate supply curve, unit # 1 will provide all required 100 MW to meet that load while unit # 2 will provide nothing. Given the network and using any power flow software, the flows on the branches can be easily computed. This is illustrated in Figure 8.2.

B: Constrained Network Case

Assume again that the network branch from bus # 1 to bus # 3 has a limit of 60 MW. For secure network operation, the flow on this branch has to be reduced by reducing the power output from unit # 1 and increasing the power output from unit # 2. As it turns out, the following dispatch (redispatch) will cause the network flow on that branch within 60 MW limit, as shown in Figure 8.3. In this case, there are two marginal units in the system.

Based on the nodal pricing method described in the previous section, the final LMPs for both unconstrained and constrained network cases at 100 MW load are shown in the Table 8.2. The GSF associated with the line 1–3 is found to be [0, −0.3333, −0.6667] for bus # 1, 2, and 3, respectively. The congestion component of LMP is found to be 0, 20, and 40 $/MWh for buses # 1, 2, and 3, respectively. Price separation among the nodes due to the congested network can be observable from the table.

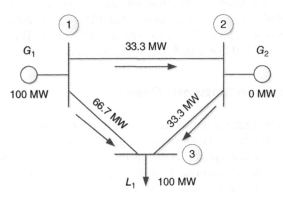

Figure 8.2 Branch Flows for the Unconstrained Network with 100 MW Load.

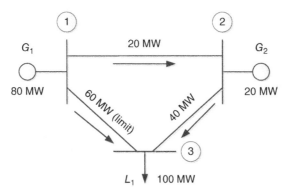

Figure 8.3   Branch Flows for the Constrained Network with 100 MW Load.

C: Constrained Network Case with Different Reference Bus

In subsections A (unconstrained network case) and B (constrained network case), we have assumed that bus # 1 is treated as the reference bus. Here, we state that the final nodal prices (LMP) will not change even if a different bus is assumed to be the reference bus. However, the components of LMP, such as MCC, will change if a different bus is used as the reference bus.

Using the same three-bus network used in the previous subsections, we assume here that the bus # 2 is the reference bus in this example. Since the generation shift factor changes if the reference bus changes, the new GSF associated with the line 1–3 is found to be [0.3333, 0, −0.3333] for bus # 1, 2, and 3, respectively. Students can reproduce the shift factors as a homework exercise. Note again that the shift factor for the reference bus (bus # 2) is zero as shift factors are computed based on the specific reference bus. We also assume that the branch 1–3 limit is 60 MW and the load at bus # 3 is 100 MW, similar in the previous example.

After using the same nodal price formulation for this example, the final LMPs are found to be 20, 40, and 60 $/MWh for bus # 1, 2, and 3, respectively, which turns out to be the same prices as in the previous example. However, the congestion component of LMP in this case is found to be −20, 0, and 20 $/MWh for bus # 1, 2, and 3, respectively, which are quite different than those in the previous example. In

Table 8.2   Nodal Prices ($/MWh) for Unconstrained and Constrained Networks for 100 MW Load

Network Status	Bus # 1	Bus # 2	Bus # 3
Unconstrained network	40	40	40
Constrained network	20	40	60

this case, the system marginal price, which is the energy price at the reference bus, is 40 $/MWh.

8.1.2.5 Co-optimization of Energy and Reserve

The recent trend in nodal pricing in the wholesale electricity markets was to co-optimize both energy and reserve (a type of ancillary services). In the real-time operation, both energy and reserve requirements have to be met simultaneously. Even though energy market and reserve market appear to be separate and independent, both energy and reserve requirements are naturally coupled. The provision of both energy and reserve can be sourced from the same generator, and thus, that generator has to make some trade-off when it faces the two markets. If not constrained by ramping capability, any potential MW of energy from a generator is a candidate for the trade-off for a potential MW of reserve provided by the same unit. The profit maximization objective of any generator will drive the unit's decision as to producing energy or providing reserve. It becomes necessary that the energy offer and reserve offer from that same generator has to be coordinated simultaneously. This condition triggers the requirement to simultaneously co-optimize both energy and reserve by jointly clearing both the energy market and reserve market in the least-bid manner. Energy-reserve co-optimization provides system operation with the modeling capability to dispatch both energy and reserve optimally and simultaneously. Energy-reserve co-optimization provides the optimal allocation of energy and reserve for each unit. The optimal solution also provides the best outcome to all dispatchable resources as well. Clearing energy and reserve in a sequential way failed to achieve the following: (1) the natural coupling effect between energy and reserve prices, (2) incentives for generating resources to follow dispatch instructions, and (3) social welfare maximization.

From the system operation point of view, the benefits of co-optimizing energy and reserve include

1. Achieving the cheapest way of meeting energy demand while maintaining the system reliability
2. Effectively determining the market-clearing prices for both energy and reserve simultaneously
3. Providing incentives for dispatch following
4. Effectively identifying units for system redispatch as well as proper compensation
5. Enhancing the reserve shortage pricing

From the market participant perspective, benefits of such co-optimization include (1) providing the optimal energy and reserve allocation that maximizes the total as-bid profit of a generating resource based on its bid-in parameters and (2) providing the optimal energy and reserve allocation that maximizes the total as-bid benefit/utility of a dispatchable load based on its bid-in parameters. The outcome of this co-optimization would be generation schedules and prices for both energy and reserve.

The objective of energy and reserve co-optimization is to minimize the total bid-based production cost (equivalent to social welfare maximization), subject to the following constraints:

1. Energy balance constraint
2. System-wide reserve constraints
3. Local reserve constraints
4. Transmission constraints
5. Resource level constraints (joint capacity, ramp, regulation, reserve capacity, etc.)

The results of this co-optimization are

1. Desired dispatch schedules
2. Nodal energy prices
3. Desired reserve schedules
4. Reserve market-clearing prices

The new OPF problem which co-optimizes both energy and reserve can be formulated as follows:

$$\text{minimize} \quad F(p^g, r) = \sum_{i \in G} c_i(p_i^g) + \sum_{j \in R} r_j * b_j \tag{8.19}$$

subject to

$$-p^g + p^d + g(\theta) = 0 \quad \longleftrightarrow \quad \lambda^e \tag{8.20}$$

$$h(\theta) \leq \bar{s} \quad \longleftrightarrow \quad \mu \tag{8.21}$$

Constraints related to the reserve market-clearing are

$$\sum_{j \in R} r_j \geq R_{\text{sys}} \quad \longleftrightarrow \quad \alpha \tag{8.22}$$

$$p_i^g - r_i \geq p_i^{g-} \quad \longleftrightarrow \quad \eta_i^- \tag{8.23}$$

$$r_i + p_i^g \leq p_i^{g+} \quad \longleftrightarrow \quad \eta_i^+ \tag{8.24}$$

$$p_i^g = 0 \quad \forall i \notin G, \tag{8.25}$$

where r_j stands for the reserve quantity offer by reserve supplier j, b_j represents the reserve price offer by reserve supplier j, $j \in R$, where R is the set of reserve suppliers participating in the reserve market which also includes a subset of generators G in the system, R_{sys} is the reserve requirement for the system, r_i represents the reserve cleared for generator i. Equations (8.23) and (8.24) represent the constraints that the sum of energy generation and reserve cleared for any participating generator should not be less than its minimum output or greater than its maximum output. The shadow price associated with system reserve requirement α represents reserve market-clearing price for the entire system. This is also equivalent to the minimum total cost increase per incremental change of the (fixed) reserve demand.

8.1.3 Uniform-Price Auction

The process of market-clearing based on either zonal or nodal pricing methods produces a market-clearing price at which the demand intersects with the supply. However, that process itself does not tell us which market participants pay or receive at which price for their consumption or production of electricity. If the market-clearing rule is based on the *uniform-price auction* and assuming there is no congestion in the system, buyers with bids at or above the clearing price pay that cleared price for the quantity purchased. Also, the suppliers with offers at or below the clearing price are paid that same price for the quantity sold. For example, a generating unit with a marginal cost of $20/MWh would be paid $50/MWh for its quantity sold in the energy market whenever the cleared price turned out to be $50/MWh. That inframarginal profit of $30/MWh is used to cover for the unit's fixed cost. Indeed, all suppliers cleared in the market are paid the highest offer of the last marginal unit among all those supplying electricity.

For the nodal pricing scheme, if there is a congestion in the system, the market-clearing price at each node in the system can become different. In this case, each consumer pays the cleared price at its consumption node and each generator receives the cleared price at its own generation node. In this situation, the market-clearing is still based on the uniform-price auction rule except the fact that the presence of a transmission congestion creates different market prices at different locations in the system.

Currently, all wholesale electricity markets in the United States employ nodal-based uniform-price auction method. In this scheme, the market operator dispatches generators in the system starting from the lowest-priced bids (such as nuclear units) and progressing to higher-priced bids (such as gas-fired generating units), until the market system has enough generation to meet consumers' demand for electricity. This process is known as *merit-order dispatch* or *economic dispatch*. Under a uniform-price auction, each generator receives the same (uniform) price based on the price of the last unit needed to meet the overall demand for electricity, regardless of each generator's bid. The bid price of the last generator used to satisfy the total demand for electricity therefore determines the wholesale price of electricity.

By analogy, most commodity markets operate with a uniform-price auction structure. As with any commodity, one unit of a product is like another unit of the product regardless of how it is produced. In the electricity market, one unit of electricity is treated the same as another unit of electricity regardless of which types of generating units produced that electricity. Some generators have lower costs and some have higher costs. However, each generator faces the same market price.

The market-clearing mechanism based on uniform-price auction is an essential feature of any electricity market that is designed to reliably provide electricity to consumers at the least-cost. The uniform-price auction plays a critical role in the least-cost dispatch of generating resources and provides an essential price signal for both short-run and long-run investment incentives. The body of economics research has supported that uniform-price auction rule increases the economic efficiency by maximizing the social welfare. In such a pricing scheme, generators are more truthful in revealing their true marginal costs by bidding as close as possible or equal to their marginal costs. The other type of pricing model is known as discriminatory pricing which is covered in the next subsection.

8.1.4 Discriminatory Pricing

In parallel with the uniform-price auction, discriminatory pricing is another type of pricing scheme. Discriminatory pricing is also known as *pay-as-bid* pricing. Discriminatory pricing is currently used in electricity markets in Britain and Iran. Under a discriminatory pricing, the prices are *not uniform* because the generators get paid for what they bid. The assumption behind adopting this discriminatory pricing in some electricity markets is that low cost supplier would bid low and hence get paid low which can translate into higher welfare for both consumers and auctioneer. This is rather a naive assumption.

Let us use one simple bidding example to understand the implication of adopting a discriminatory pricing. Assume that the marginal cost of an inexpensive supplier is $20/MWh. Under discriminatory pricing rule, the supplier would be paid $20/MWh because discriminatory pricing auction assumes that the supplier would offer close to its marginal cost even if the market-clearing price is much higher. Recall the profit maximization behavior of supplier firms. If the cleared market price is expected to be much higher, the inexpensive supplier would submit an offer which is much closer to that expected cleared price. The supplier would like to receive more revenue of course. Therefore, unless the supplying firms are forced to offer at their marginal cost, the more likely outcome would be that each supplier would offer a completely different way in a pay-as-bid environment than the uniform-price auction.

Specifically, in the pay-as-bid auction, generators would submit offers that reflected their best guess at what the cleared price will be for the most expensive needed resource, instead of bidding their actual costs as they do in a uniform-price auction. Thus, all of the bids in a pay-as-bid model would reach approximately the same level, and the cost of this auction would be essentially the same as the clearing price in a uniform-price auction. Thus, prices may not necessarily go down under pay-as-bid auction in which the generators are paid what they offer. In some situations in pay-as-bid regime, the low-cost supplier might offer much higher than the offer of a higher-cost supplier. In this case, the higher-cost supplier will be selected for dispatch, causing dispatch inefficiency.

Both theoretical and empirical work in economics research showed that both uniform-price and pay-as-bid approach result in more or less the same expected prices. Assuming there is no market power issue, the uniform-price auction yields a competitive equilibrium which is economically efficient. Social welfare is maximized in this outcome. Right amount of electricity is produced by the least-cost suppliers and this electricity is consumed by the buyers who value it the most. With expected costs equal between the two auctions, a uniform-price auction is preferred because the uniform-price auction provides transparency and ensures the selection of the least-expensive and most-efficient resources.

8.2 MODELING AND SIMULATION OF AN ELECTRICITY MARKET

The modeling and simulation of an electricity market serves many purposes. At the fundamental level, the goal of modeling and simulating an electricity market is to conduct various kinds of studies and analyses on many issues related to the electricity market. Some of the analysis goals include economic assessment of establishing an electricity market, forecast of market prices, forecast of transmission congestion, valuation of generating assets, impact of renewable energy resources, and many other types of policy analyses. Electricity market modeling can also help answer other kinds of research questions.

The modeling of an electricity market is generally complex depending on the complexity of the market. The electricity market modeling generally requires numerous "reasonable assumptions" to approximate the operations of the underlying electric system. No single model is able to replicate the exact operation of actual markets for at least two reasons: (1) difficulty in accessing the actual market data which are generally treated as confidential data and (2) the complexity of the actual electricity markets. These two factors alone make it nearly impossible to achieve an exact market model. However, we can call a market model a good one if it can capture the key attributes of a market under consideration and if it can produce reasonably good results. Therefore, modeling an electricity market is more of an art than science or engineering.

There are generally two major categories in electricity market models: (1) fundamental models and (2) market equilibrium models. It is necessary to apply different models and approaches for conducting different kinds of studies and analyses related to market, planning, policy, and research questions.

8.2.1 Fundamental Models

Fundamental models are widely used in modeling prevailing electricity markets. These models are developed to support the fundamental analysis of an electricity market. Generically, fundamental analysis refers to the analysis of various fundamental factors that affect the outcome of a market. In a financial world, the fundamental analysis is a technique that attempts to determine a security's value by focusing on the underlying factors that affect a company's actual business and its future prospects. On a broader scope, fundamental analysis can be done on industries or the economy as a whole. Fundamental models of an electricity market refer to those models that can model the key inputs of the market, such as supply, demand, network topology, and be able to reasonably replicate the inner workings of the same market. If these two steps are done right, the model should produce reasonably good results.

The first step in any market modeling is to model a *base case*. Base case generally represents the current power system with or without an operating electricity market. The key assumption that is used to differentiate these two power systems is known as *economic inefficiencies*. The power system without an electricity market is assumed to have higher economic inefficiencies while that with an electricity market is assumed to have lower economic inefficiencies. Recall that the primary goal of establishing an electricity market is to reduce or eliminate that inherent economic inefficiencies in the power system.

For the power system without an electricity market, this inefficiency is inadvertently created by individual and independent operational decisions made by various control areas trying to balance local load with local generation and power exports and imports, if any. An individual operational decision may be an optimal decision for a control area but may not be optimal if considered under a larger operational area. The same is true for operational decisions of some existing pools or ISOs. Pool-based decisions may not be optimal for a wider area that includes multiple pools. Another source of economic inefficiency is the inability of market participants to identify and realize all of the beneficial trades that might be possible. Thus, modeling a power system case with or without an electricity market requires the introduction of the right amount of economic inefficiency into the models. The difference in results, such as production cost, from these two cases generally provide the benefits of improved market efficiencies achieved by that market.

The key control variable to model such economic inefficiency is known as *hurdle rates*. The principal challenge in modeling a market structure is to find a reasonable representation of the economic inefficiencies that inevitably exist in real-world markets and, more importantly, how these inefficiencies change when moving from one market system to another. Hurdle rates, as modeled in the fundamental analysis, are used to represent the market inefficiency as a barrier to trade. These hurdle rates include real direct costs of consummating the transaction such as wheeling charges. In addition, non-cash-based barriers are also included.

Hurdle rates are comprised of three types of hurdle: import hurdle, trade hurdle, and incremental loss hurdle. Import hurdle is used to mimic the self-commitment that is the basis for current operational practices within each control area, transaction costs associated with searching out and executing bilateral trades, and other impediments to trade that bias commitment and dispatch toward internal resources. Trade hurdle reflects impediments to move power between control areas separately from the self-commitment logic embodied in the import hurdle. The trade hurdle is intended to represent both wheeling rates (cost of obtaining firm transmission) and trade impediments that become pancaked as power is wheeled across multiple control areas. Incremental loss hurdle is imposed to simulate the effect of incremental losses. They are applied to transfer out of or through a pool.

The lack of market transparency in bilateral markets prevents certain knowledge of current trading between control areas. However, the net effect of these trades is the resultant power flow between control areas. This outcome, which is known, allows a system simulation model that produces an optimized regional trading plan to be used to estimate the economic hurdle rates. Increasing the hurdle rate between two adjacent control areas will retard transactions between these areas. A trial-and-error approach, a heuristic method, can thus be used to determine what hurdle rates are necessary to approximately reproduce the observed interregional power flow.

The following example is used to illustrate the concept of hurdle rates as shown in Figure 8.4. At a point in time, let us assume that area A has a marginal cost of \$5/MWh and area B has its marginal cost of \$10/MWh. If the hurdle rate is less than \$5/MWh, power will *economically* flow from area A to area B. Area B will back down its generation and area A will increase its generation until the marginal cost of generation in area B is equal to the marginal cost of generation in area A plus the hurdle rate. If the hurdle rate, however, is more than \$5/MWh, the power will not flow from area A to area B. Economic power will not flow from area B to area A because area B has higher marginal cost than area A.

Hurdle rates for both unit commitment and dispatch decisions include all three hurdle rates mentioned above. Hurdle rates are added in both unit commitment and dispatch decisions but not necessarily of the same magnitude. Each control area

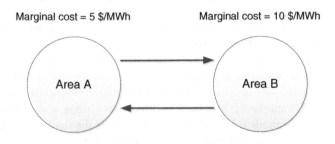

Figure 8.4   Impact of Hurdle Rate on Power Flow between Two Areas.

operator typically decides on a daily basis which generating units to commit in order to serve its load most economically with consideration of spinning reserve, thermal unit constraints, and other restrictions such as energy-limited hydro units. The hurdle rates for the commitment decision are typically set higher than those for the dispatch decision in order to reflect the control area based self-commitment that characterizes the current utility operations. Dispatch is an operational decision in real-time to meet instantaneous load with generated power in the most economic manner. Hence, the hurdle rates for dispatch decision are set lower than those that are applied to the commitment decision.

The hurdle rates for both unit commitment and dispatch are determined by trial and error for a particular system. If the hurdle rates are set too high or too low, the power flow between control areas will not match the observed flow. By adjusting the hurdle rates, the base case can be forced to replicate the historical pattern of power flow between control areas.

It is also important that the base case model replicates the historical generation pattern assuming the base case of interest is the power system without an electricity market. In order to achieve this, a number of base case simulations are run to determine the hurdle rates that result in modeled generation patterns that closely match historical generation patterns while also matching interregional power transfers. Since the change in production cost is the primary measure of immediate economic benefit, the calibration must focus on those units that have the potential to create large difference in generating costs. Many types of units will have very similar operation with, or without the market. For example, the annual generating levels for conventional hydro units will be the same for all cases. Similarly, nuclear unit operations will be the same for all cases since these base-load units will not be marginal generators under either case. Changes in production cost between two cases can only result from the substitution of lower cost generation for higher cost generation. This is except for systems in which either nuclear (France) or hydro (Brazil) resources are dominant. In general, the largest production cost differences will result from the substitution of lower cost coal-fired generation for higher cost gas-fired generation assuming gas prices are higher than coal prices. Hence, the calibration efforts should be focused on the operational patterns of coal-fired and gas-fired units which turned out to be marginal units in many electricity markets. Other systems can have different types of generation with different cost characteristics. For these systems, changes in production cost can result from the substitution among different types of low-cost and high-cost generations.

One of the fundamental models is a multiarea production costing algorithm model that incorporates detailed representations of the entire transmission system. It is based on the security-constrained unit commitment and economic dispatch algorithm. It is essentially a chronological simulation of the system operation while recognizing the transmission constraints and other unit operating constraints such as minimum up/down time. It assumes marginal cost bidding, performs a least-cost dispatch subject to thermal and contingency constraints, and calculates hourly, location-based

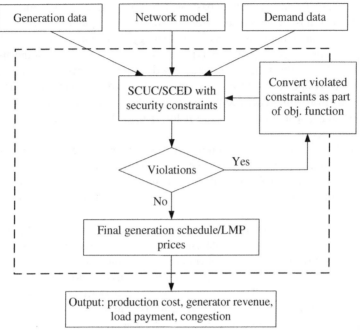

Figure 8.5 Market Simulation Process for a Nodal Market.

marginal pricing for electricity. Marginal cost bidding can be replaced by more sophisticated bidding patterns. One such process for market simulation for a nodal market is shown in Figure 8.5.

The other case, which we call *market-based model* assumes that the power system under consideration already has an operating electricity market. Thus, the market-based model requires the simulation of the power system operation which occurs in a competitive market environment. In other words, the simulation of the power system should mimic the market operation, using similar unit commitment and economic dispatch algorithm. To that end, the market simulation part of economic analysis uses the security-constrained unit commitment (SCUC) and security-constrained economic dispatch (SCED) algorithm, which is at the core of market operation, determination of LMP, and congestion pricing. There are a number of commercially available softwares in the industry that can model the fundamental model that can be used to study many kinds of issues.

8.2.2 Market Equilibrium Models

As seen in the fundamental models, electricity market modeling at least requires the representation of the underlying technical system characteristics, the physical assets and their limitations. Pure economic or financial models applicable in other contexts are not suitable for modeling market participant behavior which is an essential part of modeling an electricity market. The better model should have at least a detailed representation of the physical network system and possibly include modeling of firms' behavior. The fundamental models presented in the previous section generally do not include the modeling of firms' behavior. This is in contrast to the equilibrium market modeling.

Market equilibrium models attempt to represent the overall market behavior by taking into consideration of competition among all market participants. In other words, equilibrium models consider the simultaneous profit maximization problems of all firms competing in the market. Equilibrium models are based on the formal definition of the equilibrium which results in a set of algebraic and/or differential equations. Sometimes, these sets of *differential-algebraic equations* (DAE) are very hard to solve.

The market equilibrium models are based on game theory concepts. Electricity markets are neither pure monopoly nor perfect competition. Many believe that electricity markets are best represented by imperfect competition. The strong assumption behind these market equilibrium models was that markets are characterized by oligopolistic behavior in which a few firms dominate the market. Such imperfectly competitive market models are more complex to represent. Two prevailing models are *Cournot* model and *supply function equilibrium* (*SFE*) models. In Cournot model, firms compete based on the quantity which is treated as a decision variable. In SFE model, the decision variables of firms include both quantity and price. Although represented in a limited manner, network models are also included in these models to account for congestion pricing. Cournot model is more extensively used because it is more flexible and tractable. Even though the strategic decision variables used in both Cournot and SEF models are different, Nash equilibrium concept is the pillar for these models. In Nash equilibrium, no firm is willing to deviate from the best strategies which are the best responses to the best strategies of all other competing firms. By this way, the market reaches an equilibrium state. Market equilibrium models are also used to analyze firms' behavior and market power exercise. They can also be used in long-term planning. These models are also used to answer questions on market design.

Optimization models are a special class of market equilibrium models. In optimization models, the focus is on the behavior of one single firm which is competing in the market with the remaining firms. Optimization-based models are formulated as a single optimization problem in which one firm pursues to maximize its profit. There is a single objective function to be optimized subject to a set of technical and economic constraints. These models are suitable for developing daily bidding curves for the supplying firm in the context of an electricity market. Advances in optimization methods can enable us to represent the market equilibrium model and provide solutions that would reflect the market equilibrium condition. These methods are described next.

8.2.2.1 Mathematical Program with Equilibrium Constraints

Mathematical program with equilibrium constraints (MPEC) is a constrained non-linear programming problem in which a set of constraints are defined by a variational inequality. In this problem, two nested problems can be distinguished: the optimization problem or upper-level problem, and the variational inequality problem or lower-level problem. The importance of MPEC is based on the fact that the problems related to economics and electricity markets described in the previous sections can be modeled using MPEC. Generally, MPEC is given by

$$\text{minimize} \quad f(x, y) \tag{8.26}$$

subject to

$$g(x, y) \in Z$$
$$y \in S(x),$$

where $S(x)$ are parametric constraints, parametrized by x, on variable y. It is the solution of a mixed complementarity problem (MCP) defined by $F(x, .)$ and the bound set $\Omega(x)$. Vector $y \in S(x)$ if and only if y is an element of $\Omega(x)$ that satisfies

$$F(x, y)^T(v - y) \geq 0 \quad v \in \Omega(x) \tag{8.27}$$

MPEC is a hard problem to solve for the following reasons: (1) the lower-level complementarity constraint is generally nonconvex, (2) it has an open feasible set, (3) function $S(x)$ can have more than one element per variable x, and may not be differentiable. The MPEC formulation represents several practical cases depending on how $F(x, .)$ is represented. For example, $F(x, .)$ can be gradient with respect to variable y and another function ϕ.

Despite the inherent difficulty associated with bi-level problems, several methods have been developed based on implicit enumeration, branch and bound, penalty methods, and decomposition methods. Also, there are bi-level algorithms for the convex case. Recently, additional algorithms have been developed for larger-scale systems, where iterative algorithms are proposed based on a penalty interior-point algorithm (PIPA) implicit programming and sequential quadratic programming (SQP).

One particular application of MPEC is the analysis of competitive electricity markets based on game theory, when the optimization problem of one leader-many followers is analyzed as a Stackelberg game which is represented as a mathematical program with equilibrium constraints. The Stackelberg game assumes one leader firm after whom the remaining firms follow. The leader agent is optimized subject to the expected followers' behavior and the follower agents act in a Nash equilibrium manner. In addition, all agents are assumed to be price takers with respect to supply and demand balance.

There are some standard algorithms based on nonlinear programming (NLP) that commercial solvers use. As an illustration, we present two of these programs.

A: Equivalent Nonlinear Programming Formulation

One of the simplest ways is to represent the MPEC problem as a pure NLP problem:

$$\text{minimize} \quad f(z) \tag{8.28}$$

subject to

$$g(z) \leq 0$$
$$h(z) = 0$$
$$G(z) \geq 0$$
$$H(z) \geq 0$$
$$G(z)H(z) \leq 0$$

Then, a sequential quadratic programming (SQP) method can be used to solve the NLP problem.

B: Penalization

Following the main principles of penalization methodology, the complementarity constraints can be added to the original objective function. Then, the mathematical problem can be represented as follows:

$$\text{minimize} \quad f(z) + \lambda \sum_{i=1}^{m} G_i(z)H_i(z) \tag{8.29}$$

subject to

$$g(z) \leq 0$$
$$h(z) = 0$$
$$G(z) \geq 0$$
$$H(z) \geq 0$$

C: Example

Let us consider a bi-level optimization problem in which the main objective is to maximize the profit of a generator participant:

$$\text{maximize} \quad \textbf{(Generator profit)} \tag{8.30}$$

subject to

Generator constraints

$$\text{maximize} \quad \textbf{(Social welfare)} \tag{8.31}$$

subject to

Market-clearing price

This bi-level problem can be mathematically formulated as

$$\text{maximize} \quad f(x, y)$$

subject to

$$g(x, y) \geq 0$$
$$h(x, y) = 0$$
$$y \quad \text{solves} \begin{bmatrix} \text{minimize} \quad F(x, y) \\ \text{subject to} \\ H(x, y) = 0; \quad G(x, y) \geq 0 \end{bmatrix}$$

If the lower-level optimization problem can be linearized,

$$\nabla_y F(x, y) - \nabla_y G(x, y)^T \mu + \nabla_y H(x, y)^T \lambda = 0 H(x, y) = 0; \quad 0 \leq G(x, y)\mu = 0$$

where μ and λ are the Lagrange multipliers or shadow prices. Then, the problem can be set as a single level problem and represented using a MPEC model:

$$\text{maximize} \quad f(x, y)$$

subject to

$$g(x, y) \geq 0$$
$$h(x, y) = 0$$
$$\nabla_y F(x, y) - \nabla_y G(x, y)^T \mu + \nabla_y H(x, y)^T \lambda = 0$$
$$H(x, y) = 0; \quad 0 \leq G(x, y)\mu = 0$$

8.2.2.2 Equilibrium Problems with Equilibrium Constraints

Equilibrium problems with equilibrium constraints (EPEC) is a type of mathematical programs that is present in engineering and economics applications. One example is the case in which there are multileader–follower games in economics. And each of the leader–follower game is a Stackelberg game problem which is solved as a MPEC problem. Considering the MPEC model described in the previous section, the joint solution of all the MPEC of all the agents constitutes an EPEC problem. In other words, the EPEC problem is a collection of multiple MPEC problems in which the optimal solution is found in a simultaneous manner. Therefore, in deregulated electricity markets, EPEC model is very useful to study the strategic behavior of the multiple agents.

From the example C presented in Section 8.2.2.1, let us assume N agents, then, the corresponding EPEC is the game considering the summation of all MPEC problems described in previous section, from 1 to N. The most popular numerical methods developed to solve EPEC problems are based on diagonalization algorithms. These methods do not guarantee the global convergence. Another alternative method is based on *sequential nonlinear complementary problems* (NCP). These techniques have improved recently and become alternative methods for both MPEC and EPEC problems.

8.3 FURTHER DISCUSSIONS

Economists believe that markets are the best approach to allocate scarce resources in a society. The most critical outcome of any market is the market-clearing price (market equilibrium price). Market prices contain the information regarding consumers needs and wants to allocate resources to where consumers want them. If consumers are willing and able to pay a market price that enables suppliers to make a profit, it tells suppliers that they are using resources correctly. If the suppliers are losing money, it provides another signal. In the context of an electricity market, uniform-price auction provides far better results than discriminatory pricing in terms of dispatch efficiency, and economic efficiency.

Spot market prices go up and down due to the continuous balance of supply–demand and other factors. When prices are high, profits for suppliers will be high. This will attract more entry of new suppliers. With additional supply, prices and profits will fall. When prices are low, no additional supply will enter the market due to low or no profits. This will cause the price and profits to rise due to the tight supply. On average, spot market price is just about right and reflects *long-run average cost* (LRAC). Two factors can cause some distortions in an efficient market: (1) high barrier to entry and (2) high subsidies to electricity suppliers.

Modeling and simulation of an electricity market is an integral part of a tool kit that a researcher or a practitioner definitely needs. The changes to the design or structure of a new market regime will be costly if not done right. Modeling and simulation of a particular market design of interest can help alleviate this cost before its actual implementation. Many other questions related to policy, economic, and regulatory issues can be answered by extensive use of modeling and simulation capabilities.

CHAPTER END PROBLEMS

8.1 What are the advantages and disadvantages of adopting a zonal pricing scheme?

8.2 What are the advantages and disadvantages of adopting a nodal pricing scheme?

8.3 Why is the uniform-price mechanism widely used in many electricity markets around the world, compared with discriminatory pricing?

8.4 In the example of three-bus network system, compute the nodal prices (LMPs) if the load level at bus # 3 is 150 MW and the branch limit from bus # 1 to # 3 is 70 MW.

8.5 In the example of three-bus network system, compute the nodal prices (LMPs) if the load level at bus # 3 is 100 MW and the branch limit from bus # 1 to # 3 is 50 MW.

8.6 A five-bus network is given in the Figure 8.6. There are two generators in the system, with one generator located at bus # 1 (unit # 1) and another generator at bus # 5 (unit # 2) with demands at bus # 2 (200 MW), # 3 (100 MW), and # 4 (300 MW). assume that the generator at bus # 1 is a cheap coal-fired unit with the marginal cost of $40/MW with capacity of 400 MW. The generator at bus # 5 (unit # 2) is an expensive gas-fired unit with the marginal cost of $80/MW with capacity of 600 MW. Bus # 3 is assumed to be the reference bus.

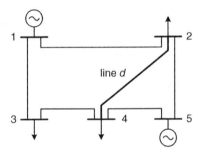

Figure 8.6 Five Bus System.

a. For the unconstrained-network case, determine the total generation cost and flow on the line d;

b. if the flow limit on line d is 60 MW, what is the total generation cost;

c. determine the final LMPs at each node in the system if shift factors on line d by each node is given as 0.1818, 0.3636, 0.0000, −0.1818, and 0.0909, respectively.

8.7 Select a practical electricity market of your choice and analyze their market prices (day-ahead or real-time energy markets) for a historical year. What did you observe? What can you tell about why prices are that level?

8.8 Formulate an alternative OPF problem in which price-responsive demands and reserve market participation by consumers are also included in the objective function, in addition to generator offers into both energy and reserve markets. Try to include relevant equality and inequality constraints in the formulation.

FURTHER READING

1. Bjørndal M, Jørnsten K. Zonal pricing in a deregulated electricity market. *The Energy Journal* 2001;22(1):51–73.

2. Hogan WW. Nodes and zones in electricity markets: seeking simplified congestion pricing. In: *Designing Competitive Electricity Markets*, Chao HP, Huntington HG, editors. Kluwer Academic Publishers; 1998. Chapter 3, pp. 33–62.

3. Schweppe FC, Caramanis MC, Tabors RD, Bohn RE. *Spot Pricing of Electricity*. Boston: Kluwer Academic Publishers; 1988.

4. Hogan WW. Contract networks for electric power transmission. *Journal of Regulatory Economics* 1992;4:211–242.

5. Chao HP, Peck S. A market mechanism for electric power transmission. *Journal of Regulatory Economics* 1996;10:25–59.

6. Chen L, Suzuki H, Wachi T, Shimura Y. Components of nodal prices for electric power systems. *IEEE Transactions on Power Systems* 2002;17(1):41–49.

7. Rivier M, Pérez-Arriaga IJ. Computation and decomposition of spot prices for transmission pricing. In: Proceedings of 11th PSC Conference, 1993.

8. Wood AJ, Wollenberg BF. *Power Generation, Operation, and Control*, 2nd edition. New York: John Wiley & Sons; 1996.

9. Tierney SF, Schatzki T, Mukerji R. Uniform-pricing versus pay-as-bid in wholesale electricity markets: does it make a difference? Report by Analysis Group and NYISO, March 2008.

10. Cramton P, Stoft S. The convergence of market designs for adequate generating capacity. White Paper for the California Electricity Oversight Board, 2006.

11. Cramton P, Stoft S. Uniform-price auctions in electricity markets. Mimeo. March 2006.

12. Lin Y, Jordan G, Zhu J, Sanford M, Babcock WH. Economic analysis of establishing regional transmission organization and standard market design in the Southeast. *IEEE Transactions on Power Systems* 2006;21(4):1520–1527.

13. Lin J. Market-based transmission planning model in PJM electricity market. In: Proceedings of 6th International Conference on European Energy Market, May 2009.

14. Lin J. Modeling and simulation of PJM and Northeastern RTOs for interregional planning. In: Proceedings of IEEE Power Engineering Society General Meeting, July 2013.

15. Lin J. Impact analysis of Entergy– MISO integration on power system economics. In: Proceedings of IEEE Power Engineering Society General Meeting, July 2014.

16. Ventosa M, Baillo A, Ramos A, Rivier M. Electricity market modeling trends. *Energy Policy* 2005;33(7):897–913.

17. Smeers Y. Computable equilibrium models and the restructuring of the European electricity and gas markets. *Energy Journal* 1997;18(4):1–31.

18. Hobbs BF. Linear complementarity models of Nash– Cournot competition in bilateral and POOLCO power markets. *IEEE Transactions on Power Systems* 2001;16(2):194–202.

19. Day CJ, Hobbs BF, Pang JS. Oligopolistic competition in power networks: a conjectured supply function approach. *IEEE Transactions on Power Systems* 2002;17(3):597–607.

20. Klemperer PD, Meyer MA. Supply function equilibria in oligopoly under uncertainty. *Econometrica* 1989;57(6):1243–1277.

21. Rudkevich A, Duckworth M, Rosen R. Modeling electricity pricing in a deregulated generation industry: the potential for oligopoly pricing in a Poolco. *Energy Journal* 1998;19(3):19–48.

22. Gabriel SA, Conejo AJ, Fuller JD, Hobbs BF, Ruiz C. *Complementarity Modeling in Energy Markets*. New York: Springer; 2013.

23. Ehrenmann A. Equilibrium problems with equilibrium constraints and their applications to electricity markets. Ph.D. dissertation, Fitzwilliam College, Cambridge University, 2004.

Chapter 9

Evaluation of an Electricity Market

This chapter will cover important issues related to and methods used in evaluating an electricity market to judge whether a particular market of interest is competitive and functioning well. The chapter will emphasize on presenting materials in a more fundamental manner as a theoretical basis, rather than actually judging the merits of existing electricity markets. In this regard, the standard way, perhaps a golden rule, in assessing an electricity market is to use the structure–conduct–performance model. Assessment of markets includes assessment of both energy and capacity markets as well as other related markets. In energy markets, it is necessary to assess, at least estimate, whether the participating generators are making sufficient level of revenues so as to maintain their long-term economic viability. In a nodal market, the issue of revenue adequacy for FTR is also important. How can we judge that a capacity market is functioning as planned. These topics will be explored in this Chapter.

9.1 MARKET COMPETITIVENESS

The first question that must be asked in evaluating an electricity market is whether the market is competitive or not. Economists believe that the social welfare is maximized *only* when the market is competitive. If the market competition is weak, social welfare would suffer particularly for consumers. With this reasoning, the fundamental goal of establishing an electricity market is to make it competitive. The fundamental theories about market and market competitiveness are treated in Chapter 3. Recall that a competitive market is a market with a sufficient number of both buyers and sellers such that no one buyer or seller is able to influence or exercise control over the market or the price. This definition of a competitive market also applies to electricity

Electricity Markets: Theories and Applications, First Edition. Jeremy Lin and Fernando H. Magnago.
© 2017 by The Institute of Electrical and Electronics Engineers, Inc. Published 2017 by John Wiley & Sons, Inc.

markets. An ability to exercise control over the market or the price is related to the fundamental issue of market power, as previously discussed. Therefore, evaluating whether a market is competitive is about evaluating whether the market is conducive to the existence and exercise of market power.

9.2 MARKET POWER

Using the definition used by antitrust agencies in the United States, market power is *the ability of an individual supplier or group of suppliers to profitably maintain prices above competitive levels for a significant period of time*. In a power system, a significant period of time can be as short as several dispatch periods in the presence of price spikes. Simply put, market power reduces the competition among suppliers and thus makes the market less competitive. Dealing with market power is the central theme of this chapter.

As noted previously, the key components of supply chain in the power industry are generation, transmission, distribution, and delivery of electricity to consumers. The issue of market power related just to a particular component of this supply chain, such as generation, is known as *horizontal market power*. For example, a company tries to own a large number of generators in a market to gain a significant market share. This company can have significant horizontal market power. Another type of market power is known as *vertical market power*. Assume that a company is involved in more than one activity, such as both generation and transmission, in which the company attempts to increase its profit in one segment by using the dominant position in another segment.

As electricity markets are getting more sophisticated, so is the market manipulation as another form of market abuse. For example, in a nodal market, a company can artificially create transmission congestion in one area to benefit its own generators on the constrained side of the system. Other types of market manipulation are possible.

9.3 STRUCTURE–CONDUCT–PERFORMANCE MODEL

The starting point in analyzing markets and industries is to employ an analytical framework known as *structure, conduct, and performance paradigm*. It is a pillar of industrial organization theory and was developed in 1959 by Economist Joe S Bain Jr. The main idea in this paradigm is that an industry performance is determined by the conduct of the firms within the boundaries of that industry which in turn depend on the structure of the market.

9.3.1 Structure

The structure of a market includes a set of variables that are relatively stable over time and affect the behavior of sellers and/or buyers. The way in which markets do or do not follow a perfect competition model, depends largely on supply concentration, demand concentration, product differentiation, and barrier to entry. For example, a market may not be competitive if a small group of firms controls the majority of the supply (supply concentration) or if the barrier to entry is high. The structure of the market is also determined by the product nature and available technology. The unique properties and characteristics of electricity and different methods of energy production must be properly considered in evaluating the structure of an electricity market.

The competitiveness of a market structure is evaluated based on a number of criteria. The common metrics used in assessing a market structure are measure of market share, m-firm concentration index, and estimate of *Herfindahl–Hirschman index* (HHI). These metrics are also used in econometric and antitrust analysis as a potential indicator of market power.

9.3.1.1 Market Share
Market share is one of the concentration indices used to measure the supplier concentration of a market. The assumption here is that the higher the supplier concentration, the more likely that these suppliers are able to exercise market power. The definition of market share is simply the percentage of market share held by any firm i in the market. Firm i's market share can be written as

$$\beta_i = \frac{q_i}{Q},\tag{9.1}$$

where $(i = 1, ..., F)$ and $\sum_{i=1}^{F} \beta_i = 1$, β_i is the market share of firm i, F is the total number of firms in the market/industry, q_i is the amount/size of the product quantity held by firm i, and Q is the total product quantity in the market.

When using the market share index, it is important to clearly define the market product and market boundaries. It is certainly not a straightforward exercise in applying this metric in different market products of an electricity market. For example, capacity and energy production are clearly two different products. Also, the energy production in this hour and the next are not easily substitutable, hence they are different products. These factors have to be carefully considered before the market share index is used.

9.3.1.2 m-Firm Concentration Ratio
A concept related to market share is the *m-firm concentration ratio* which is a measure of combined market share of a given number of large firms in a market or industry. This ratio can be used to measure the degree of market concentration. The m-firm concentration ratio has the following formula:

$$R_m = \sum_{i=1}^{m} \beta_i, \quad \text{where} \quad m \le F,\tag{9.2}$$

where R_m is the concentration ratio of m largest firms, β_i is the market share of firm i, and F is the total number of firms in the market/industry.

For example, the four-firm concentration ratio provides the combined market share of the largest four firms in an industry, as a proportion of the total size of the industry. The higher the ratio, the more concentrated the industry, and the greater the potential for exercise of market power.

9.3.1.3 Herfindahl–Hirschman Index

Herfindahl–Hirschman index or *HHI* is a well-known structural index used to detect market concentration. It is a commonly accepted metric in assessing the degree of concentration—from non-concentrated to moderately to highly concentrated—of a particular market. The underlying assumption is that a market is *more* competitive if it is *less* concentrated. On the other hand, a market is *less* competitive if it is *moderately* to *highly* concentrated. The HHI has the following formula:

$$ \text{HHI} = \sum_{i=1}^{F} (\beta_i)^2, \tag{9.3} $$

where β_i is the market share of each firm i competing in the market and F is the total number of firms in that market.

A numerical example is given to illustrate the concept of HHI. Assume a particular market consists of five firms with market shares of 10, 20, 20, 20, and 30%. Then, the HHI index for this market is 2200. The HHI takes into account the relative size distribution of the firms in the market. If the market is controlled by a single firm, the HHI value will be 10,000 (pure monopoly). The HHI value becomes smaller, or even approach to zero (atomistic market) if there are many competing firms with relatively equal size. The U.S. Department of Justice (DOJ) and Federal Trade Commission (FTC) provided guidelines in interpreting HHI values. Based on these guidelines, markets are considered highly concentrated if the HHI is in excess of 2500 points and moderately concentrated if the HHI is between 1500 and 2500 points. Markets are considered unconcentrated if the HHI is below 1500 points. Quantitative measures of market concentration or market power are often used in assessing the effect of horizontal mergers, as one component of competition analysis. The US competition authorities use HHI index to screen mergers, and identify those that warrant further investigation. Also, transactions that increase the HHI value by more than 200 points in highly concentrated markets are considered likely to enhance market power.

Electricity markets are unique in the sense that the aggregate market can be separated into smaller markets typically caused by transmission constraints (at least in nodal markets). In this case, the smaller markets on the constrained side of the aggregate market can become highly concentrated, making them vulnerable to market power exercise. In this situation, other measures or metrics are necessary to detect such market power that might exist in the constrained markets.

9.3.1.4 Entropy Index A lesser known index for supplier concentration is called *entropy index* (*EI*). The entropy measure has its theoretical foundations in information theory and measures the *ex ante* expected information content of a distribution. Entropy index is the negative of the sum of the market shares times their logarithmic values as shown in the following formula:

$$\text{EI} = -\sum_{i=1}^{F}(\beta_i \times log(\beta_i)), \tag{9.4}$$

where the relevant variables were defined in the previous sections. EI equals $log(F)$ when all firms have an equal share. EI is zero when only one firm controls all shares. The value of entropy varies inversely with the degree of concentration. Thus, decreasing values of EI indicate higher levels of concentration. In contrast to HHI, more weights are given to smaller firms and a smaller weight is attached to larger market shares.

The concentration metrics—market share, m-firm concentration ratio, HHI, and entropy index—are based on measuring the supply side of the market only. Additional measures were developed to include the demand side of the market in measuring the potential market power. These measures of detecting structural market power include pivotal supplier test, residual supply index, and residual demand analysis.

9.3.1.5 Pivotal Supplier Test Pivotal supplier test is a test to measure a structural market power. The key feature of a pivotal supplier test is to consider both supply and demand conditions in measuring market power. It can be used in both energy and capacity markets. The test measures whether a given generator is necessary or pivotal in serving a demand at a given time. Specifically, it tries to examine whether the capacity of a generator is larger than the surplus supply (the difference between total supply and total demand) in the wholesale market. A supplier may be pivotal at a system and/or a local level. A supplier is considered pivotal if its capacity is needed to satisfy system or local level requirements. In other words, the system/local requirements cannot be met after the removal of the supplier's capacity. In this case, the supplier is considered pivotal. For a local market, a generation owner or a group of generation owners is pivotal if the output of the owners' generation facilities is required in order to relieve a transmission constraint. When a generation owner or a group of generation owners is pivotal, it has the ability to increase the market price above the competitive level.

A numerical example below will help illustrate the concept of pivotal supplier test. Assume that the demand of a particular electricity market at a particular market period is 1000 MW. There are five suppliers with the following capacity offers, shown in Table 9.1.

Table 9.1   Capacity Offer by Market Suppliers

Suppliers	Offer MW
S-1	150
S-2	200
S-3	400
S-4	300
S-5	200

If the offer MW capacity of all suppliers are accounted, the total offer capacity would be 1250 MW, which exceeds the required demand. The necessary steps to do the pivotal supplier test for a particular supplier are as follows:

1. Compute the total offer MW capacity of all suppliers.
2. Subtract the MW capacity of the supplier of interest from the total offer MW capacity.
3. Recompute the total offer capacity of remaining suppliers.
4. Determine if the supplier of interest is pivotal.

In step-4, a supplier is considered pivotal if the recomputed total offer capacity (from step-3) is less than the required demand. On the flip side, a supplier is not pivotal if the recomputed total offer capacity exceeds the required demand. In the given numerical example, suppliers 3 and 4 are pivotal.

Another way to determine if a supplier is pivotal is to find the surplus supply which is the difference between total supply and total demand. In the given example, this surplus supply is 250 MW. A supplier is pivotal if its capacity exceeds that surplus supply. Again, suppliers 3 and 4 are pivotal because their individual offer capacity exceeds that surplus supply. For these pivotal suppliers, some sort of market power mitigation schemes are needed to address the market power issue.

In the given example of the pivotal supplier test, we are only concerned with one supplier at a time, particularly the largest supplier. Generally, the largest supplier or suppliers are more likely to be pivotal. In certain circumstances, it would be worthwhile to conduct pivotal supplier test for more than one supplier in a joint manner. The following are natural extensions of applying pivotal supplier test for more than one supplier simultaneously:

1. **Two Pivotal Supplier Test**: This test requires the comparison of the required demand with the remaining supply capacity after removing the combined capacity of the two largest suppliers. The two largest suppliers are pivotal if the remaining supply capacity after their removal is not sufficient to meet the demand. Otherwise, they are not pivotal. Another possible application of the

test is in determining if one largest supplier and another supplier, not necessarily the second largest, are jointly pivotal. Conceptually, the combination of any two suppliers of any size can be tested to determine if they are jointly pivotal.

2. **Three Pivotal Supplier Test**: Similar to the two pivotal supplier test, instead of removing the combined capacity of just the two largest suppliers, the combined capacity of the three largest suppliers are removed in determining the remaining supply capacity. The three largest suppliers are pivotal if the remaining supply capacity is less than the required demand. Three pivotal supplier test explicitly incorporates the impact of the excess supply and implicitly accounts for the impact of the price elasticity of demand in the market power tests. In some practical markets, three pivotal supplier test is usually employed when a particular market area is transmission-constrained. In this case, the three suppliers to be tested in the constrained area include two largest suppliers and another supplier which may hold market power.

Several existing wholesale electricity markets in the United States, such as PJM, ERCOT, and Cal-ISO, employ some sort of pivotal supplier screening tests as a trigger for identification and mitigation of market power. In addition, FERC uses pivotal supplier test, along with other tests, to determine if a generation company is pivotal or not before making decision to grant or deny the company in charging market-based rates for energy.

9.3.1.6 Residual Supply Index The *residual supply index* (RSI), developed by Cal-ISO, measures the aggregate capacity of all suppliers except one supplier, typically the largest supplier, as a fraction of total demand. The RSI for the supplier *j* has the following formula:

$$
\text{RSI}_j = \frac{\sum_{i=1}^{F} S_i - S_j}{\sum D},
\tag{9.5}
$$

where S_i is the capacity of each supplier i, F is the total number of suppliers including supplier j, S_j is the capacity of supplier j, and $\sum D$ is the total demand of the system.

The supplier j is pivotal when the RSI_j is less than one because the capacity of this supplier is needed to meet the demand. If RSI is greater than one, the supplier j has little or no influence over the market-clearing price because the remaining supply, hence named *residual supply*, has more than enough capacities to meet the demand. By intuition, the supplier with the lowest RSI among other suppliers in a market is the supplier with the largest capacity. Lower values of RSI can be interpreted as yielding greater market power for these suppliers of interest. Cal-ISO has used this metric extensively and found the strong correlation between hourly RSI and hourly price-cost markup in the California market. Empirical studies showed that a threshold level of RSI for any given period to avoid loss of load probability is 1.1.

9.3.1.7 Residual Demand Analysis   The residual demand curve is the individual firm's demand curve which is the portion of market demand that is not supplied by other firms in the market. Thus, it is the market demand function minus the quantity supplied by other firms at each price. Mathematically, it can be written as

$$D_i(p) = D(p) - S_{-i}(p), \tag{9.6}$$

where $D_i(p)$ is the residual demand curve faced by the firm i, $D(p)$ is the market demand curve, and $S_{-i}(p)$ is quantity supplied by all firms except firm i.

The assumption behind using this residual demand analysis is that a firm's market power is largely determined by the firm-specific demand curve so that demand curve estimation can provide useful insight into the firm's market power. The market theory holds that in a competitive market, a firm faces a highly *elastic* residual demand curve and therefore has little or no ability to raise market price above the competitive level. On the other hand, in a non-competitive market in which the firm is pivotal, the firm faces a highly *inelastic* residual demand curve. In this case, the firm can raise the price without losing much on sales. In other words, the extent to which a firm can exercise market power by raising the price depends on the elasticity of residual demand faced by that firm. Hence, residual demand analysis includes the analysis of both residual demand and residual demand elasticity faced by a firm or many firms.

9.3.2   Conduct

The conduct represents the way in which buyers and sellers behave, both among themselves and among each other. The behavior of firms includes choosing their own strategic initiatives, investments in research, advertising levels, and even collusions. The conduct that is of interest here is about the firm's conduct related to the actual exercise of market power. Recall that the firm's goal is to maximize its profit. The profit is the difference between how much a firm earns and how much it costs for the firm to produce the electricity. One rational way to increase the profit is by raising the market price. In the case of an electricity market, supplier firms try to raise their offers significantly higher than competitive levels or hope that other supplier firms will do the same so as to achieve a higher market price. Market price in an electricity market is the price cleared by the market operator. General increase of offers by many suppliers will shift the supply curve to the left, which will increase the cleared market price higher than it would have been if the suppliers offer at the competitive levels. In a zonal pricing market, the generators will receive the zonal clearing price (assuming uniform-price auction). In a nodal pricing market, the generators will receive the cleared nodal price at the bus at which the generator is located.

Therefore, in investigating the behavior of supplier (or buying) firms in an electricity market, it is important to examine the supplier's offers (price and quantity), buyer's bids (price and quantity), and revenues it receives or payment it makes in any relevant product market of an electricity market. Detailed knowledge about the specific features of the specific product market is also necessary to judge whether the strategic behavior of a particular firm is detrimental to the market competitiveness. For example, in a particular nodal market which is experiencing a binding transmission constraint, a non-pivotal supplier located within the network-constrained area may possess significant market power under that network condition. Tacit collusion among participating generators is a real possibility and must be thoroughly analyzed. However, it may be difficult to prove as the agreements are implicit. The following indices are generally used to determine if a supplier has potential for market power exercise.

9.3.2.1 Lerner Index
The *Lerner index (LI)*, developed in 1934 by an Economist named Abba Lerner, describes a firm's market power by measuring the extent to which a given firm's prices exceed its marginal costs. It is measured as the difference between the price of a good or service and its marginal cost, expressed as a fraction of the price. The Lerner index has the following formula:

$$LI = \frac{P - MC}{P}, \tag{9.7}$$

where P is the price of a good or service, for example, price of electricity in an electricity market, that the firm receives, MC is the marginal cost of the firm. We assume here that $P > MC$.

The key variable in LI is the market price. If the demand (buyer firms) is more responsive to the price (i.e., elastic demand), prices may not go as high as compared to a situation in which demand is inelastic. In a sense, the elastic demand is a useful mechanism to counteract the potential existence and exercise of market power by supplier firms.

The higher the Lerner index, the greater the markup in price over the marginal cost. Higher values of Lerner index also indicate greater market power. However, a higher LI does not necessarily mean that the firm in question is exercising market power. Prices may exceed the firm's marginal cost for a variety of legitimate reasons. The value of Lerner index can rise for the following reasons: (1) the firm's market share increases, (2) market price elasticity of demand falls, and (3) competitors' supply elasticity falls.

Firms also need to earn above and beyond their marginal costs to cover for the high fixed costs. The difficulty in estimating the Lerner index is attributable to the difficulty in calculating a firm's marginal cost of production. Practically, the marginal costs of firms are difficult to calculate. For this reason, modified Lerner index is sometimes used instead of the original Lerner index. The modified Lerner index (LI_2) has the following form:

$$LI_2 = \frac{P - MC}{P} = \frac{1}{\epsilon} = \frac{S}{\epsilon_M + (1 - S)f}, \tag{9.8}$$

where P is the firm's price, MC is the firm's marginal or incremental cost, S is the firm's market share, ϵ is the firm price elasticity of demand, ϵ_M is the market price elasticity of demand, and f is the supply elasticity of the competitive fringe.

9.3.3 Performance

Several metrics are typically used for assessing the performance of different electricity markets. For example, the performance metrics for assessing energy markets include measures and analysis of market prices, markup by generators, price convergence between day-ahead, intraday and real-time markets, and whether there is a scarcity condition during the assessment period. Prices are one of the key outcomes of markets. Price is an indicator of the level of competition in a market. In a competitive electricity market, prices are directly related to the marginal cost of the marginal unit that is required to serve the last increment of load for any given period. Prices are generally related to the supply and demand conditions of the market and thus illustrates the relationship between the price elasticity of demand and price. Measures of market prices can be done hourly, daily, monthly, and yearly. Statistical measures of prices including average, minimum, maximum, and standard deviation values are also useful metrics.

The performance metrics used for a capacity market are similar to those used in the energy markets. Thus, performance metrics for assessing capacity markets include measures of market prices and revenues received by the generators. Similar performance metrics can be applied in assessing other product markets of an electricity market. The overall performance of an electricity market can also be measured by comparing the results of firms in the industry in efficiency terms. Different ratios can be used to assess different profitability levels of supplier firms. While there are many attributes and variables that are used in assessing the overall performance of an electricity market, it is worthwhile to describe the assessment of generator net revenue, FTR revenue adequacy, and performance of a capacity market.

9.3.3.1 Analysis of Generator Net Revenue
Analysis of net revenues earned by supplying firms in an electricity market is an important part of assessing the overall health of the market. Supplier firms are represented by different types of generating resources. Generally, the types of these resources include nuclear, coal-fired steam, gas-fired combined cycles and combustion turbine, solar, wind, and

other types. When compared to annualized fixed costs, net revenue is an indicator of generation investment profitability, and thus is an important measure of overall market performance as well as a measure of the incentive to invest in new generation (barrier to entry) to serve in electricity markets.

By definition, the net revenue equals the total revenues earned less the variable cost of energy production for each generator. In other words, net revenue is the amount that remains, after subtracting short-run variable costs of energy production from gross revenues, to cover the fixed costs, which include a return on investment, depreciation, taxes and fixed operation, and maintenance expenses. The net revenues of generators for the future can be estimated by comparing the forecasted revenues with forecasted total variable production costs. On a high level, the net revenue of a generator i has the following form:

$$\pi_i = \sum R_i - \sum C_i \qquad \forall i \in G, \tag{9.9}$$

where π_i is the net revenue of a generator i; G is the set of generators in a market; $\sum R_i$ is the total revenues that a generator i receives from all possible sources including energy markets, and represented by $\sum R_i = (R_e + R_c + R_{as} + R_o)$, where R_e is revenue received from energy market, R_c is revenue received from capacity market, R_{as} is revenue received from ancillary service markets, and R_o is revenue received from other service provisions; $\sum C_i$ is the total variable production cost and represented by $\sum C_i = (C_f + C_{om} + C_e + C_o)$, where C_f is fuel cost, C_{om} is operational and maintenance cost, C_e is emission cost, and C_o is other variable costs.

In energy-only markets, the generators can earn revenues from energy market, ancillary service markets (if they exist) and from the provision of black start and reactive services. In energy-capacity markets, the generators can earn revenues from energy market, capacity market, ancillary service markets, and other service provisions. These revenues are primarily determined by the final market-clearing prices in each respective market. For most generators, the variable production cost is largely determined by the price of fuel that the generators burn to generate electricity. Identification and estimation of these fuel prices is a critical part of this analysis. These prices can vary from one period to the next, which can cause significant variations and fluctuations in net revenues earned by generators.

Given the same variable production cost, higher market prices from electricity markets would translate into higher net revenues for generators. On the other hand, given the same revenues from various electricity markets, higher input cost, such as higher fuel cost, can reduce the net revenues. Extreme events in the system can cause larger fluctuations in net revenues.

In the long-run equilibrium of a perfectly competitive energy-only market, net revenue from the energy market would be expected to equal the total of all annualized fixed costs for the marginal unit, including a competitive return on investment. In the long-run equilibrium of a perfectly competitive energy-capacity market, net revenue from all sources including energy, capacity, and ancillary service payments, would be expected to equal the annualized fixed costs of generation for the marginal unit.

Net revenue is a measure of whether generators are receiving competitive returns on invested capital and of whether market prices are high enough to encourage entry of new capacities. In actual wholesale power markets in which equilibrium seldom occurs, net revenue is expected to fluctuate above and below the equilibrium level based on the actual conditions of all relevant markets.

9.3.3.2 Revenue Adequacy of FTR In the nodal electricity markets in the United States, the FTR market product is also introduced to allow the market participants to be able to hedge against the risks associated with volatile market prices. Market prices fluctuate not only due to fluctuating fuel prices but also due to transmission congestions that occur within the system. Recall the three components of the nodal price. Generally, given the fixed system marginal price, market prices go higher on the congested side of the constraint in the market system. Since power typically flows in the direction from generators to load consumers, the congestion is typically in that direction, causing the market prices at load customers higher than the prices at the generators. Reversal of power flow and resultant congestions is entirely possible at different system operating conditions.

When FTR products were first introduced, there was also an assumption that when FTRs are sold to the market participants, they will receive the full funding. This is known as the principle of FTR revenue adequacy. FTR revenue adequacy means that the revenues collected from the economic dispatch in the form of congestion payments are sufficient to fully fund the payments for all the FTRs held by FTR holders. The theory holds that for a given grid configuration, if FTRs are simultaneously feasible, the FTR funding will be revenue adequate from security-constrained economic dispatch with nodal prices for any given load and generation levels. The rationale for this theory is that economic dispatch is feasible, and must be at least as valuable as many other feasible solutions. In other words, the economic dispatch in the day-ahead market must be at least as valuable as the FTR-implied feasible dispatch, valued at the locational prices. Hence, there must be enough economic surplus value to buy out all the FTRs and reconfigure the pattern of flows according to the economic dispatch.

However, this theory is seriously challenged when it is applied to the real-world markets. There are a number of possible reasons for experiencing FTR revenue inadequacy. For example, the network status at the time when FTRs are auctioned off and the network used in the day-ahead market clearing will be different, possibly due to transmission upgrades and unexpected outages of existing transmission lines for some period. This posed the problem of infeasibility for some FTRs in real-time, while these FTRs are awarded earlier based on the network model at that time. The uncertainty of network status also increases for the longer time horizon compared with shorter time horizon.

One of the key steps in initial auctioning of FTRs is called *simultaneous feasibility test* (*SFT*). It is a test to ensure that all subscribed transmission entitlements can be supported and guaranteed by the capability of the existing transmission system. It is a test to ensure the energy market is revenue adequate under normal system conditions. It is neither a system reliability test nor intended to model actual system conditions. When SFT was run at the initial stage of FTR auctions, the following key input are used in the process:

1. Uncompensated parallel flow injections (loop flow)
2. Known transmission outages
3. Existing FTRs
4. Facility ratings
5. Network model
6. List of contingencies
7. Reactive interface ratings (if any)

Consequently, if the values of any of these variables at the time of FTR auctions deviate from those in the day-ahead market clearing, this can create an issue of FTR revenue inadequacy. In other words, any of these input parameters are possible sources for less than full funding of FTRs.

FTR underfunding can increase the overall uncertainty of the market and reduction in market efficiency. For example, FTR revenue inadequacy can cause significant uncertainty regarding forward FTR values. It can also undermine the hedging and planning functionality of FTRs. Electricity markets that avoided the problem of FTR revenue inadequacy would enjoy higher liquidity and efficiency in their markets. Possible remedies to resolve the issue of FTR underfunding include conservative allocation of FTRs at the initial auction stage to account for the future network uncertainty. Other measures can be implemented to resolve this problem.

9.3.3.3 Performance of a Capacity Market
The topic of capacity market was treated in an earlier chapter. The capacity market is designed to send the long-term price signals necessary for building additional power supply resources needed to meet the predicted energy demand in the future so as to ensure long-term grid reliability. In other words, the given capacity market design must ensure in providing the appropriate incentives for suppliers to make investments to address availability concerns and to ensure that capacity is reliable.

In a scarcity condition, capacity availability is particularly important. The capacity unavailability in the scarcity condition would defeat the purpose of having a capacity market. One should consider the option of redesigning the capacity market if the capacity market is not producing the desired results.

Using the paradigm of structure–conduct–performance, the first step in assessing a capacity market is to ensure that the capacity market is structurally competitive. Second, it must be ensured that no market participants (suppliers) exercise market power. If they do, some kind of market power mitigation must be put in place. Third, the outcome of the capacity market prices must be competitive.

In the end, capacity reliability is critical for ultimate system reliability, lack of which will be detrimental to the final consumers. The costs of cleared prices from the capacity market are ultimately borne by electric consumers because resource reliability is beneficial to them.

On a high level, the first question to ask about the performance of a capacity market is whether it attracts sufficient level of new generations as well as retaining existing generations. Just because a capacity resource is cleared in a capacity market does not necessarily mean that this capacity resource will always be available in the intended delivery year. In fact, market clearing is one thing and capacity resource performance is another. That is why some capacity markets use some sort of penalty scheme associated with the non-performance of the generators. One common approach is that a penalty is imposed to a capacity resource which is not available when it is supposed to be available. Alignment between incentives and penalties related to the performance of generators is expected to be optimal.

Market power issue in the capacity market is also important. It must be ensured that no individual supplier or a group of suppliers have any influence on the capacity market and the final market outcome.

9.4 OTHER MARKET POWER ISSUES

Electricity markets that exist today are generally complex. Thus, other forms of market power abuses are possible. In fact, such abuses did occur in real-world electricity markets. One of such abuses is known as *affiliate abuse*.

For the electricity markets in the United States, transmission business is regulated while generation business is subject to participation in the competitive markets. Due to strict federal rules, the probability of a collusion between a transmission company and its affiliated generation company under the same holding company (vertical market power) is extremely low. However, the collusion among generation companies or financial companies which participate in competitive electricity markets (horizontal market power) is a real possibility. For example, a parent financial company would own two or more affiliated financial companies. On the surface, these companies may seem to be acting independently. In reality, one or more such affiliated companies may be colluding, thus benefiting one company by the action of another affiliated one.

9.5 FURTHER DISCUSSIONS

Electricity markets are evolving, so as the metrics and measurements that are used to assess such evolving markets. For example, electricity markets are dynamic, therefore, those metrics should be adaptable and applicable for such dynamic markets. For markets in which transmission constraints are considered, measures of market power under the transmission-constrained condition should take into account of this unique condition. Authors believe that this is still an active area of research.

As electricity markets are complex, detecting market power and marker power abuse in a practical setting is much more broader in scope than the assessment metrics mentioned in this chapter. Lawsuits and litigations also become part of the reality because the stakes in terms of financial rewards and penalty are relatively high.

CHAPTER END PROBLEMS

9.1   What are the limitations of using market share as a structural index for an electricity market?

9.2   What are the limitations of using HHI as a structural index for detecting market power exercise in an electricity market? Support your answer with empirical evidence?

9.3   Can you find a product in an electricity market in which the structurally competitive market with competitive behavior by market participants is producing non-competitive market outcome (performance)?

9.4   Compute the HHI values for the two markets: (i) Market-A—there are seven firms in this market with market shares of 5, 5, 10, 15, 20, 20, and 25%; (ii) Market-B—there are only three firms in this market with market shares of 30, 30, and 40%. What can you tell about these markets?

9.5   The market shares of generating companies in an electricity market are given by 10, 15, 15, 20, 20, and 20% . Two generating companies decide to merge. What is the increase in HHI from pre-merger to post-merger status? Pick a different set of two companies that plan to merge. How much does the post-merger HHI change from pre-merger? What can you observe?

9.6   The energy demand of a transmission-constrained load-pocket market at a particular market period is 10,000 MW. The offer capacities of the suppliers within that load-pocket area are given in Table 9.2. Determine which supplier (or suppliers) is pivotal in meeting the required demand. Ignore the import capability and import power into the area. What measures should be taken if there is a pivotal supplier in this particular market?

Table 9.2   Capacity Offer by Market Suppliers

Suppliers	Offer MW
S-1	1500
S-2	2000
S-3	2000
S-4	1000
S-5	1000
S-6	2000
S-7	3000

9.7 What are the advantages and disadvantages of using residual demand analysis as a method to detect market power?

9.8 Identify an industry or market in which high supplier concentration leads to high profits for the firms. Support your answer with empirical evidence.

9.9 Show that the average Lerner index is equal to the HHI index divided by the elasticity of demand under Cournot competition.

9.10 What are the reasons for HHI becoming more popular index than entropy index in its application to market power detection in electricity markets?

FURTHER READING

1. *Horizontal Merger Guidelines*. U.S. Department of Justice and the Federal Trade Commission. Available online at https://www.justice.gov/

2. Twomey P, Green R, Neuhoff K, Newbery D. A review of the monitoring of market power: the possible roles of TSOs in monitoring for market power issues in congested transmission systems. MIT Center for Energy and Environmental Policy Research, 05-002 WP, March 2005.

3. Baker JB, Bresnahan TF. Empirical methods of identifying and measuring market power. *Antitrust Law Journal* 1992;61(1):3–16.

4. Bain JS. *Industrial Organization*, 2nd edition. New York: John Wiley & Sons; 1968.

5. Bushnell J, Day C, Duckworth M, Green R, Halseth A, Read EG, Rogers JS, Rudkevich A, Scott T, Smeers Y, and Huntington H. An international comparison of models for measuring market power in electricity. EMF Working Paper 17.1, Energy Modeling Forum, Stanford University, 1999.

6. Sheffrin A. Predicting market power using the residual supply index. Presented to FERC Market Monitoring Workshop, December 3–4, 2002.

7. Sheffrin A. Empirical evidence of strategic bidding in California ISO real-time market. In: *Electric Pricing in Transition*. Norwell, MA: Kluwer; 2002. pp. 267–281.

8. Wolak FA. An empirical analysis of the impact of hedge contracts on bidding behavior in a competitive electricity market. *International Economic Journal* 2000;14(2):1–39.

9. Wolak FA. Measuring unilateral market power in wholesale electricity markets: the California market 1998–2000. *The American Economic Review* 2003;93(2):425–430.

10. Lerner AP. The concept of monopoly and the measurement of monopoly power. *The Review of Economic Studies* 1934;1(3):157–175.

11. PJM state of the market. Monitoring Analytics. [Online]. http://www.monitoringanalytics.com/home/index.shtml

12. Hogan WW. Financial transmission rights, revenue adequacy and multi-settlement electricity markets. March 2013. [Online]. http://www.hks.harvard.edu/fs/whogan/Hogan_FTR_Rev_Adequacy_031813.pdf

13. California ISO. [Online]. http://www.caiso.com/market/Pages/TransmissionOperations/Default.aspx

Chapter 10

Transmission Planning Under Electricity Market Regime

This chapter will focus on transmission planning, particularly in the competitive electricity market environment. Various drivers for conducting transmission planning will be described. Both reliability and market-based transmission planning will be addressed in more detail, including various required analyses and methodologies. Transmission planning that crosses the borders of two or more adjacent transmission market systems will be presented with a workable methodology to facilitate a process in which a cross-border transmission project can be proposed, selected, and eventually built. Finally, we will present a transmission planning in an electricity market environment and beyond that has moved toward a more competitive environment in which nonincumbent transmission developers can compete with the incumbent transmission owners on a level playing field. Even though the transmission planning in this chapter is largely drawn from US experiences, the relevant analyses and methodologies are equally applicable to other market regimes throughout the world.

10.1 OVERVIEW

There are two major functions in a power system: *planning* and *operation*. The topic of power system operation is treated in earlier chapters. Readers can refer to these chapters and other excellent references for understanding about power system operation. Generally, power system planning can be classified into three groups: (1) generation system planning, (2) transmission system planning, and (3) distribution system planning. Generation system planning is briefly treated below. Transmission system planning under the electricity market regime is the main topic of this chapter. Distribution system planning is outside the scope of this book.

Electricity Markets: Theories and Applications, First Edition. Jeremy Lin and Fernando H. Magnago.
© 2017 by The Institute of Electrical and Electronics Engineers, Inc. Published 2017 by John Wiley & Sons, Inc.

Under the traditional utility model, an electric utility company is vertically integrated, sometimes known as vertically integrated utility. A single utility owns and plans all sectors of the supply chain: generation, transmission, and distribution. In some ways, the integrated planning of both generation and transmission systems is also known as *integrated resource planning* (IRP). However, in those regions in which electricity markets were developed as part of the power industry restructuring since the 1990s, the generation and transmission businesses were separated, or unbundled. Under the restructured environment, generation and transmission assets are owned by different businesses or entities and thus are planned, operated, and managed by separate companies or subsidiaries of the same company. Therefore, under electricity market regime, the generation expansion is dependent on the business decisions of the individual merchant generation developers or generation owners as the generation sector was open for competition.

Based on the market incentives provided via market prices, for-profit seeking merchant generation developers typically decide the location, size, and type of generators to be interconnected into the system. Once the proposal to build a new generating resource is made, the *independent system planner* (ISP) conducts necessary technical studies such as feasibility, impact, and facilities studies to determine if each proposed generator to be interconnected into the system causes any violations against acceptable reliability criteria. If there are reliability-related violations triggered by the proposed generator, the appropriate transmission network upgrades are determined and associated costs for the network upgrades are assigned to the respective generators. During the process, the generation developer has a choice to interconnect the generator to the grid or not. The developer can move forward with the process of interconnecting the generator by paying appropriate fees for the network upgrades. If there are no reliability-related issues due to the interconnection of the proposed generator, the generation developer can move forward with the interconnection process. At any point in time, the developer can decide to stop pursuing the generation interconnection process. In this case, the proposed generator will drop out of the queue process.

In contrast to the generation expansion planning, the transmission system planning under the electricity market regime took a different path. Recall that transmission business is still regulated under the restructured industry because it is a natural monopoly. Instead of relying on the multiple decisions of individual generation developers as in the case of generation expansion, the planning of the transmission system is organized as a centralized activity administered by a centralized planning authority or entity. In this case, the centralized planning entity is the independent system planner (ISP). The fundamental objective of transmission system planning is to develop the transmission system as economically as possible so as to maintain an acceptable level of system reliability. Maintenance of an acceptable level of reliability for a transmission system is done in two major steps: (1) evaluate the transmission system based on the industry-acceptable reliability standards and criteria and (2) recommend transmission solutions in areas where there are reliability-related violations now and the future. The transmission system development is generally associated with determination of reinforcement alternatives, additions, improvements, and their implementation time. A decision on the retirement or replacement of aging system equipment is also an important task in transmission planning.

By FERC requirement, one of the eight minimum functions of an RTO is "Planning and Expansion." To carry out this function, an RTO must consider all possible alternatives to meet the future load while supporting the competitive market. These alternatives are not just confined to transmission-based solutions. These alternatives should also include generation-based solutions (merchant generation, using either conventional generation technology, dispersed generation or renewable generation), energy efficiency/conservation approaches, and even demand response programs. Based on the economics or reliability reason, an RTO will find one alternative or another to be the best option for planning and expansion of the transmission system and support of the market. Among these alternatives, we will focus on the transmission-based solution in this chapter.

10.2 KEY DRIVERS FOR TRANSMISSION PLANNING

Using the *regional transmission expansion plan* (RTEP) of PJM as an example, a schematic of the key drivers for transmission system expansion is shown in Figure 10.1. As can be seen in the figure, the expansion of transmission system can be driven by many factors. The list of factors presented here is certainly not exhaustive. These factors will be further explained.

10.2.1 Load Forecast

Load growth is one of the key drivers for system reinforcement for both generation and transmission. Thus, an accurate load forecast is an essential input to the transmission planning process. The purpose of load forecast is to predict a load for the system at a time or series of times in the future. In addition to modeling transmission topology and generation, the forecasted load values are typically used as key input in the system planning studies.

Load forecasts can be divided into three categories based on the future time horizon of interest:

1. **Short-Term**: Typically from 1 hour to 1 week
2. **Medium-Term**: Usually from 1 week to 1 year
3. **Long-Term**: Longer than 1 year

While the shorter-term load forecasts primarily rely on weather forecasts, the longer-term load forecasts depend on other macroeconomic factors such as population growth, economic development. Generally, only medium-term and longer-term load forecasts are applicable to system planning studies.

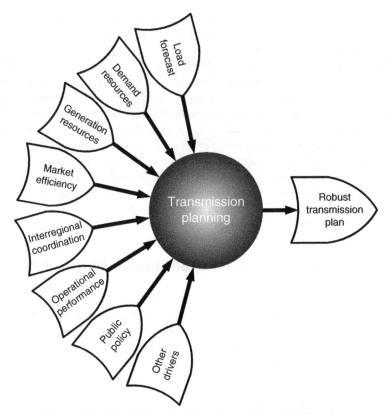

Figure 10.1 Key Drivers for Transmission Expansion Planning.

Most forecasting methods use statistical techniques or artificial intelligence algorithms such as regression, neural networks, fuzzy logic, and expert systems. Two of the methods, so-called end-use and econometric approaches are broadly used for medium-term and long-term load forecasting. A variety of methods including similar day approach, various regression models, time series, neural networks, statistical learning algorithms, fuzzy logic, and expert systems, have been developed for short-term load forecasting. Regression analysis can be linear, nonlinear, univariate, or multivariate. Probabilistic time series can be used in place of deterministic time series method. The accuracy of load forecasting depends not only on the load forecasting techniques, but also on the accuracy of input parameters, such as forecasted weather scenarios, into the model.

10.2.2 Demand Resources

Demand resources or demand-side resources are closely related to the load forecast as these resources can affect the forecasted or final realized load. Nowadays, there are various options and approaches available to load consumers to better manage their load profiles and peak loads. Collectively, these resources are called *demand response resources* (DRRs) or *demand response* (DR).

Demand response can be defined as any resource that has the capability to change or reduce the electricity consumption at a given time. The mode to change the electricity consumption can be instantaneous or pre-scheduled. Since DR is a demand-side resource, in contrast to the supply-side resource, the key players of DR resources are those who consume, not supply, electricity. Typically, DR players are represented by residential, commercial, and industrial customers of electricity. DR aggregators can also represent the aggregated demands for a group of smaller residential customers. DR can include peak-shaving, time-of-use, load management, and price-based demand response. Price-based DR encourages customers to voluntarily schedule electricity consumptions based on price signals. This is in contrast to interruptible-load programs which are managed and controlled by the electric utilities. With DRs, the load consumers or DR aggregators have full control over their load consumptions.

With the development of electricity markets, DR providers can also participate in energy, capacity, and other markets. The participation and clearing of DRs in the capacity market not only effectively change the peak load requirement of the system but also defer the need for certain expensive generators. The increased participation of DRs in the system will lower the load (peak and energy) to serve, and thus can defer the requirement for transmission expansion as well. Varying amount of DRs can also affect the power flow pattern in the system. The system operator can also use DRs to manage supply scarcity condition and possibly avoid a potential blackout.

10.2.3 Generation Resources

The reliability of the whole power system largely depends on the reliability of the power generation system. The generation in the system should be such that it can supply and meet the demand at all times under the condition of random outages (or forced outages) and planned maintenance outages. Generation planning means managing the retiring generators and expanding new generators to meet the uncertainly growing load. Under the traditional utility paradigm, generation planning is typically done by a single electric utility for its operating area.

Under the restructured market environment, the additions of the new generations and the retirements of the old generations in the system are left to the business decisions of the generator developers and owners. Retirement of generation resources can be due to equipment aging, economic reasons or federal, and state environmental policies. Recall that some electricity markets are energy-only markets, while others are energy-capacity markets. Under either market, the independent system planner (ISP) must ensure there is a sufficient level of generation to meet the load plus reserve margin for foreseeable future. Since generation business is open for competition, the merchant developers or other generator owners can propose new generations to be interconnected into the system.

ISP also conducts interconnection studies which assess the potential system impacts caused by new generation, merchant transmission, long-term firm, and other transmission service needs. The key steps in interconnection studies for new generators include feasibility, system impact, and facility planning studies which identify transmission upgrades needed to ensure that new resources can be interconnected to the grid without violating industry-established reliability standards. The interconnection of smaller units less than 20 MW such as wind or solar powered projects can be expedited by a streamlined study process.

A rapidly shifting capacity mix, triggered by many factors, may require numerous transmission upgrades to ensure system reliability. The reasons for this shifting capacity mix include new capacity powered by several shale natural gas discoveries and tens of thousands of MW of announced coal-fired generator deactivations due to stricter environmental regulations. The impact of *renewable portfolio standards* (RPS) is also another factor. Transmission planning requires modeling of generation additions and retirements as accurately as possible.

10.2.4 Market Efficiency

Market efficiency driver is fairly new to the transmission planning process. The objective of this driver is to identify transmission bottlenecks in the system caused by binding congestions and to develop transmission solutions that would relieve those congestions. By adding such transmission solutions into the system, the market efficiency will improve and hence social welfare will increase. Market efficiency driver is applicable only to a transmission system based on which an electricity market operates. Particularly, the operation of the electricity market should be based on nodal pricing scheme because the internal congestions can be explicitly identified and quantified in the nodal-based market regime. Market efficiency model is also known as economic model of market-based transmission planning.

Market efficiency studies entail production cost analysis in evaluating transmission enhancements for their economic value based on their ability to relieve persistent congestions in the market system. The salient feature of this model is that it is based on the forecasted congestion of the system for future years, based on the best possible system topology and generation scenario, while meeting the current reliability criteria with minimum reliable reserve margin. Production cost analysis used in the market efficiency studies apply the security-constrained unit commitment and economic dispatch (SCUC/ED) algorithm which is at the heart of determining nodal prices (LMP) and valuable congestion information. The goal of this analysis is to mimic the actual market operation.

Economic objective of market efficiency analysis is to compare the benefits and costs of proposed transmission projects will using reasonable estimate of discount rates and the *net present value* (NPV) method so as to determine the economic viability of the transmission upgrades. Each ISP in the United States uses a slightly different set of metrics in estimating the economic benefits that proposed transmission projects will bring into the system. The metrics used for measuring economic benefits generally include a single or composite measure of reduction (saving) in production cost, load payment, and congestion. The benefit-to-cost threshold of 1.25 is typically used in screening among many proposed projects. Those projects with benefit-to-cost ratios that equal or exceed that threshold are further assessed for any additional system impacts. Sensitivity analyses are done to determine the various key drivers for influencing the economic outcome. The detailed procedure for conducting a market efficiency analysis is further described in a later section.

10.2.5 Interregional Coordination

Most power systems are interconnected. Proposed or planned transmission projects at or near the border of one system can impact the neighboring system. Therefore, transmission planning at or near the border of a system must be done by the ISP in coordination with the potentially affected neighboring systems. This is the role of interregional or cross-border transmission planning. Due to a recent push by the US federal regulatory authority via Order 1000, it becomes necessary to consider transmission planning (upgrades or new transmission facilities) that touches more than one regional entity.

Under the restructured environment, cross-border transmission planning must be done not just to improve system reliability but also to reduce cross-border congestions. Due to the convergence of some key market rules across contiguous electricity markets and various economic activities, inter-RTO/ISO trading activities have increased. This phenomenon causes more power transfers across adjacent electricity markets and sometimes creates cross-border congestions. Assuming that these power transfers are based on the free enterprise mentality, the seemingly frictional borders appear to create some economic barriers to the freedom of these power transfers. In this case, transmission planning or expansion should be done across multiple-market system areas with the goal of reducing or even eliminating cross-border congestions. Some established markets such as PJM and MISO already put in place some formal procedures to deal with cross-border congestions on daily basis. However, future upgrades or expansion of cross-border transmission is inevitably a better solution to actually reduce or eliminate such congestions.

In addition to the joint cross-border transmission planning, coordinations among neighboring systems provide a process in dealing with, in joint manner, queued interconnection projects, market-to-market congestions and public policy mandates such as RPS that can have impact on more than one system. However, conducting a cross-border transmission planning across neighboring systems is also a very time-consuming process as it requires coordination, cooperation, and agreement among different parties. Different operating rules and protocols among neighboring systems can hinder or complicate the joint planning process. Despite these obstacles, ISPs have to find practical ways to deal with these hurdles and effectively conduct cross-border transmission expansion to reap the benefits associated with these expansion plans.

Interregional or cross-border coordination between or among neighboring systems will become more important in the near future, especially when planning authorities of different regions or different power systems need to interconnect with each other for various reasons.

10.2.6 Operational Performance

There are many possible system conditions that can have significant impact on the operational performance of the system. Some of the typical operating conditions of interest include

1. Unusually high peak load during very hot summer
2. Significant generation outages during extremely cold winter
3. Fuel shortage for a significant number of natural gas-fired generators
4. High-voltage conditions during light load periods
5. Transmission loading relief (TLR)
6. Post-contingency local load relief warning (PCLLRW) events
7. Persistent uplift payments in the market
8. Extreme weather events, such as tsunami or super-storm which can cause severe damages to power system components

These operational issues can cause real power shortage, reactive power shortage, voltage problems (high-voltage/low-voltage), and other system-related problems. During any of these events, the system operator attempts to resolve these issues by using available operational tools or options at their disposal. Transmission switching such as opening/closing a line or multiple lines, injecting or withdrawing reactive power from reactive power compensation devices, managing real power generation and demand response, changing tap positions of on-load tap-changing (OLTC) transformers, and phase shifters are some of the available control knobs for the system operator. Despite these efforts, a few of the operational problems continue to persist. In those particular situations, the better solution would be simply adding a new transmission line or upgrading an existing line or lines which can effectively resolve such persistent operational problems. In general, transmission upgrades or expansions can be an effective remedy in resolving some of the operational issues.

10.2.7 Public Policy

Public policy is also a relatively new driver in the system planning arena. Public policy requirements are defined as enacted statutes and regulations promulgated by a relevant jurisdiction, whether within a state or at the federal level. These policy initiatives are driven by other exogenous factors such as global warming, technological improvements, and changing human behavior.

For example, many states in the United States have issued laws and policy directives which require that certain percentage of energy generation in their respective states be sourced from *renewable energy sources* (RES). Such policies are known as renewable portfolio standards (RPS). While there may be some variations in terms of which specific resources will meet the definition of renewable energy resources in each state, it is generally agreed that the majority of those renewable energy will come from energy produced by solar and wind energy resources. Other types of renewable energy resources, such as geothermal energy and biomass, may play some minor roles.

The status of RPS in each state of the United States as of August 2016 is shown in Figure 10.2. For example, the state of California has one of the most aggressive renewable energy goals compared with many other states in the country, requiring that about 50% of energy by the year 2030 be sourced from renewable energy resources. Similarly, the state of Illinois has renewable energy goals of meeting 25% of energy from RES by the year 2026. In contrast, there are other states which have less aggressive renewable energy goals. For example, the state of Ohio has mandated that 12.5% of energy be sourced from RES by the year 2026.

These policy mandates will likely increase the percentage share of renewable energy resources in the generation mix in the respective jurisdictions. Significant amount of RES can have a significant impact on the system and can cause a significant strain on the transmission system. In some instances, significant amount of new transmissions or upgrades will be necessary in order for the system to reliably deliver the renewable energy to the customer load. Therefore, policy makers, system planners, and other key stakeholders have keen interest in understanding the impact of the large-scale integration of wind and other renewables—often distant from the load centers they serve—and thus conducted numerous system studies including transmission need assessments and potential solutions. Thus, public policy issue becomes another important driver for conducting transmission planning.

10.2.8 Other Drivers

There are a few other drivers which can trigger the new transmission development. For example, aging infrastructure is one of these drivers. Every equipment in the power system has a useful technical life. At or near the end of its useful life, the equipment has to be removed or replaced with new one because continued operation of the old equipment poses some risks of negative impact on the reliability of the transmission system. A new transmission line will be needed just to replace another line which reaches its end of life. Industry guidelines indicate equipment life standards such as wood structures (35–55 years), conductor and connectors (40–60 years), and porcelain insulators (50 years).

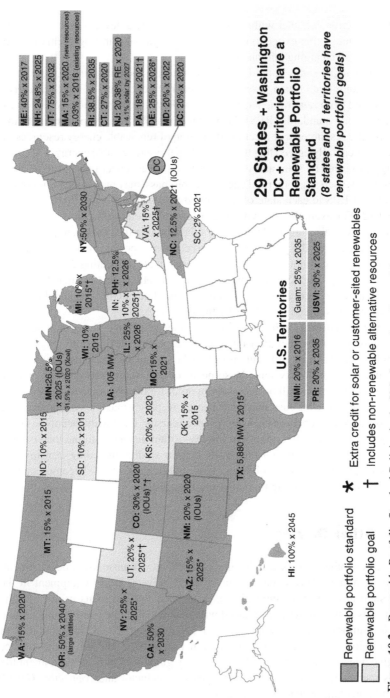

Figure 10.2 Renewable Portfolio Standard Policies in the U.S. Source: http://www.dsireusa.org/resources/detailed-summary-maps/ Public domain.

10.3 RELIABILITY-BASED TRANSMISSION PLANNING

Reliability-based transmission planning entails expansion or upgrades of transmission facilities based on reliability analyses which evaluate thermal, voltage, transient stability, and short circuit performance, per industry-established reliability planning criteria. Similarly, analyses are performed to resolve operational performance issues encountered in real-time operation.

10.3.1 Establishing a Reference System

An ISP or any entity responsible for system planning needs to assess their relevant transmission system to judge whether the system can accommodate forecast demand, committed resources, and commitments for firm transmission service for a specified time frame in accordance with industry and other applicable reliability and design standards. Areas not in compliance with the standards are identified and enhancement plans to achieve compliance have to be developed.

Reliability analysis can be done on an annual basis or a continuous basis or as needed. In order to establish a reference point of the system for any subsequent reliability analyses, it is necessary to conduct a baseline analysis of system adequacy and security. The purposes of this analysis are threefold:

1. To identify areas of the system, as planned, which are not in compliance with applicable industry and other applicable regional reliability standards. The baseline system is analyzed using the same criteria and analysis methods that are used for assessing the impact of proposed new interconnection projects. This ensures that the need for system enhancements due to baseline system requirements and those enhancements due to new projects are determined in a consistent and equitable manner.

2. To identify and recommend facility enhancement plans, including cost estimates and estimated in-service dates, to bring those areas into compliance.

3. To establish the baseline facilities and costs for system reliability. This forms the reference point for determining facilities and expansion costs for interconnections to the transmission system that cause the need for facilities beyond those required for system reliability. The "baseline" analysis and the resulting expansion plans serve as the base system for conducting feasibility studies for all proposed generation and/or merchant transmission facility interconnection projects and subsequent system impact studies.

10.3.2 Reliability Analysis of a Reference System

The most fundamental responsibility of an ISP is to plan a safe and reliable transmission system that serves all long-term firm transmission uses on a comparable and not unduly discriminatory basis. This responsibility is addressed by conducting reliability-based transmission planning. Reliability-based transmission planning consists of a series of detailed analyses that ensure reliability under the most stringent of the applicable regional or local criteria. To accomplish this, the transmission expansion plan has to be updated on an annual or a continuous basis for the future time horizon of interest. This planning process entails several steps in which each step has its assumptions, processes, and criteria.

Reliability planning can involve a near-term and a longer term review. Different ISPs choose different time horizons for near-term and longer term depending on many factors. For example, the near-term analysis can be applicable for the current year through the current year plus 5. The longer term review can be applicable for the current year plus 6 through plus 15 or longer. Each transmission plan review entails multiple analysis steps subject to the specific criteria that depend on the specific facilities and the type of analysis being performed. Each ISP has its own approval process for recommended transmission upgrades and facilities. Generally, the independent board of directors of an ISP has the ultimate authority in approving the recommended transmission plans. Other planning authorities may have different customs and procedures for such approvals.

10.3.3 Near-Term Reliability Review

The near-term reliability review provides reinforcement for criteria violations that are revealed by applicable contingency and other analyses. The limits of network facilities used in the analysis should be established consistent with the requirements of applicable standards. System conditions revealed as near violations should be monitored and remedied as needed in the following year near-term analysis. Violations that occur in many deliverability areas or severe violations in any one area can be referred to the long-term analysis for added study of possible more robust system enhancement. Assuming the near-term analysis is from the current year through the current year plus 5, this system review shall include system peak load for either year one or year two, and for year five.

For the annual evaluation of the near-term, sensitivity cases can be utilized to demonstrate the impact of changes to the basic assumptions used in the model. To accomplish this, the sensitivity analysis in the planning assessment must vary one or more of the following conditions by a sufficient amount to stress the system within a range of credible conditions that demonstrate a measurable change in system response:

1. Real and reactive forecasted load
2. Expected power transfers
3. Expected in service dates of new or modified transmission facilities
4. Reactive resource capability
5. Generation additions, retirements, or other dispatch scenarios

6. Controllable loads and demand side management

7. Duration or timing of known transmission outages

Occasionally, planned generation modifications or changes in the transmission topology can trigger restudy and the issuance of a baseline addendum. Each year or as necessary, a reassessment of a reference case for all years of near-term must be done to verify the continued need for or modification of past recommended upgrades. Violations of all thermal and voltage criteria resulting from the near-term analyses are produced using solved AC power flow solutions. Initial massive contingency screening may use DC power flow solution techniques. The key steps involved in a near-term reliability review are as follows:

1. Reference system power flow case

2. Reference system thermal analysis

3. Reference system voltage analysis

4. Load deliverability analysis—thermal and voltage

5. Generation deliverability analysis—thermal

6. Reference system stability analysis

7. Other analyses

These reliability related steps must be followed by a scenario analysis that ensures the robustness of the plan by looking at the impacts of variations in key parameters selected by the ISP. Each of these steps is described in more details in the following sections.

10.3.3.1 Reference System Power Flow Case   The reference system power flow case and the analysis techniques comprise the full set of analysis assumptions and parameters for reliability analysis. The ISP must begin the process of developing the reference system power flow case by incorporating all of the current system parameters and assumptions. These assumptions include current loads, installed generating capacity, transmission and generation maintenance, system topology, and firm transactions. Respective stakeholders also review and provide feedback on the modeling and assumptions to the development of the reference power system models used to perform the reliability analyses.

For a transmission system with a capacity market auction, the results of any locational capacity market auctions are used as a subset of the amount and location of generation or demand side resources to be included in the reliability modeling. However, generation or demand side resources that either do not bid or do not clear in any locational capacity market auction are excluded in the reliability modeling.

Contingency definitions used in the reliability analysis must conform to those applicable industry-standard contingency definitions. Where the physical design of connections or breaker arrangements results in the outage of more than the faulted equipment when a fault is cleared, the additional facilities should also be taken out of service in the contingency definition. For example, if a transformer is tapped off a line without a breaker, both the line and transformer must be removed from service as a single contingency event.

Contingency definitions for double circuit tower line outages shall include any two adjacent (vertically or horizontally) circuits on a common structure, but shall exclude circuits that share a common structure for one mile or less. The loss of more than two circuits on a common structure constitutes an extreme event. The ISP must also coordinate with adjacent planning coordinators and transmission planners to ensure that contingencies on adjacent systems which may impact their system be included in the contingency list.

10.3.3.2 Reference System Thermal Analysis   Baseline thermal analysis is a thorough analysis of the reference power flow case to ensure thermal adequacy of the system based on normal (applicable to system normal conditions prior to contingencies) and emergency (applicable after the occurrence of a contingency) thermal ratings specific to the transmission facilities being examined. It is based on a load forecast, typically 50/50, from the latest available load forecast report. 50/50 load forecast means 50% probability that the actual load is higher or lower than the projected load. It encompasses an exhaustive analysis of all credible events and the most critical common mode outages. Final results are supported with AC power flow solutions.

For normal conditions, all facilities shall be loaded within their normal thermal ratings. For each single contingency, all facilities shall be loaded within their emergency thermal ratings. After each single contingency and allowing phase shifter, redispatch, and topology changes to be made, post-contingency loadings of all facilities shall be within their applicable normal thermal ratings. For the more severe contingencies, along with only transformer tap and switched shunt adjustments enabled, post-contingency loadings of all facilities shall be within their applicable emergency thermal ratings.

10.3.3.3 Reference System Voltage Analysis   Voltage analysis of the reference case parallels the thermal analysis. It uses the same power flow and examines voltage criteria for all the same credible events. Voltage criteria are also examined for compliance. Analysis will simulate the expected automatic operation of existing and planned devices designed to provide steady-state control of electrical system quantities when such devices impact the study area. Those devices may include phase-shifting transformers, load tap-changing transformers, and switched capacitors and inductors. System performances for both a voltage drop criteria (where applicable) and an absolute voltage criteria are also examined.

The voltage drop is calculated as the decrease in bus voltage from the initial steady-state power flow to the post-contingency power flow. The post-contingency power flow is solved with generators holding a local generator bus voltage to a pre-contingency level consistent with applicable specifications. In most instances, this is the pre-contingency generator bus voltage. Additionally, all phase shifters, transformer taps, switched shunts, and DC lines are locked for the post-contingency solution. Static VAR compensators (SVCs) are allowed to regulate and fast-switched capacitors are enabled.

The absolute voltage criteria is examined for the same contingency set by allowing transformer taps, switched shunts, and SVCs to regulate locking phase shifters and allowing generators to hold steady-state voltage criteria which is generally an agreed upon voltage on the high-voltage bus at the generator location.

In all instances, voltage results are observed based on applicable voltage criteria for the system or area. All voltage violations are recorded and reported and tentative solutions are developed. Post-contingency voltage analysis shall also include the impact of tripping generators where the simulated generator bus voltages or the high side of the generation step up (GSU) transformer are less than known or assumed minimum generator steady-state of ride through voltage limitations. Tentative solutions are developed for all violations.

10.3.3.4 Load Deliverability Analysis   The load deliverability tests are a unique set of analyses designed to ensure that the transmission system provides a comparable transmission function throughout the system. These tests ensure that the transmission system is adequate to deliver each load area's requirements from the aggregate of system generation. The tests develop an expected value of loading after testing an extensive array of probabilistic dispatches to determine thermal limits. A deterministic dispatch method is used to create imports for the voltage criteria test. The transmission system reliability criterion used is 1 event of failure in 25 years. This is intended to design transmission so that it is not more limiting than the generation system which is planned based on the reliability criterion of 1 failure event in 10 years.

Each load area's deliverability target transfer level to achieve the transmission reliability criterion is separately developed using a probabilistic modeling of the load and generation system. The load deliverability tests described here measure the design transfer level supported by the transmission system for comparison to the target transfer level. Transmission upgrades are specified to achieve the target transfer level as necessary.

The thermal test in the load deliverability analysis examines the deliverability under the stressed conditions of a 90/10 summer load forecast. This is a forecast that only has a 10% chance of being exceeded. The transfer limit to the load is determined for system normal and all single contingencies under ten thousand load study area dispatches with calculated probabilities of occurrence. The dispatches are developed randomly based on the available data for each generating unit. This results in an expected value of system transfer capability that is compared to the target level to determine system adequacy. All applicable normal and emergency ratings of the transmission system are applied. The steady-state and single contingency power flows are solved consistent with the similar solutions described for the thermal analyses of the reference system case. This testing procedure for voltage analysis is similar to the thermal load deliverability test except that voltage criteria are evaluated and that a deterministic dispatch procedure is used to increase study area imports. The voltage tests and criteria are the same as those performed for the voltage analyses of the reference system case.

10.3.3.5 Generation Deliverability Analysis The generation deliverability test for the reliability analysis ensures that, consistent with the load deliverability single contingency testing procedure, the transmission system is capable of delivering the aggregate system generating capacity at peak load with all firm transmission service modeled. The procedure ensures that there exists sufficient transmission capability in all areas of the system to export an amount of generation capacity at least equal to the amount of certified capacity resources in each *area*. Areas, as referred to in the generation deliverability test, are unique to each study and depend on the electrical system characteristics that may limit transfer of capacity resources. For generator deliverability, areas are defined with respect to each transmission element that may limit transfer of the aggregate of certified installed generating capacity. The cluster of generators with significant impacts on the potentially limiting element is the *area* for that element. The starting point power flow is the same power flow case set up for the reference case analysis. Thus, the same load and ratings criteria from that reference case are applicable. The same contingencies used for load deliverability test apply and the same single contingency power flow solution techniques also apply.

One additional step is applied after generation deliverability is ensured consistent with the load deliverability tests. The additional step is required by system reliability criteria that call for adequate and secure transmission during certain common mode outages. The procedure mirrors the generator deliverability procedure with single contingency with somewhat lower deliverability requirements consistent with the increased severity of the contingencies.

10.3.3.6 Reference System Stability Analysis The ISP must ensure generator and system stability during its interconnection studies for each new generator. In addition, the stability analysis must be performed for any generator that is deemed to be critical for stability issues. Reference system stability cases must be developed and stability analyses are performed based on these cases. These analyses ensure that the system is transiently stable and that all system oscillations display positive damping with applicable damping ratios. Generator stability studies are performed for critical system conditions, which include light load and peak load for three phase faults with normal clearing plus single line-to-ground faults with delayed clearing. Also, specific transmission owner designated faults are examined for plants on their respective systems. Finally, the ISP can initiate special stability studies as necessary. The trigger for such special studies commonly includes, but is not limited to, conditions arising from operational performance reviews or major equipment outages.

Additional analyses may be necessary for system reliability. These analyses include (1) light load reliability analysis, (2) winter peak reliability analysis, (3) spare equipment availability, and (4) maximum credible disturbance review.

10.3.3.7 Light Load Reliability Analysis The light load reliability analysis ensures that the transmission system is capable of delivering the system generating capacity at light load. The 50% of 50/50 summer peak demand level is typically chosen as being representative of an average light load condition. The system generating capability modeling assumption for this analysis is that the generation modeled reflects generation by fuel class that historically operates during the light load demand level.

The starting point power flow is the same power flow case set up for the reference case analysis, with adjustment to the model for the light load demand level, interchange, and accompanying generation dispatch. Adjustments are made to the study area of the model as well as areas surrounding the study area that impact loadings on facilities in the study area. Interchange levels for the various system zones reflect a statistical average of typical interchange values for off-peak hours from previous years. Load level, interchange, and generation dispatch for external areas impacting system facilities are based on statistical averages for previous off-peak periods. Thus, the same network model and criteria for the reference system apply. The flowgates ultimately used in the light load reliability analysis are determined by running all contingencies and monitoring all market monitored facilities and all *bulk electric system* (BES) facilities. The same single contingency power flow solution techniques used in other reference system reliability tests also apply.

10.3.3.8 Winter Peak Reliability Analysis The winter peak reliability analysis ensures that the transmission system is capable of delivering the system generating capacity at winter peak. The 50/50 winter peak demand level is typically chosen as being representative of a typical winter peak condition. The system generating capability modeling assumption for this analysis is that the generation modeled reflects generation by fuel class that historically operates during the winter peak demand level.

The starting point power flow is the same power flow case set that is used for the reference system analysis, with adjustments to the model for the winter peak demand level, winter peak load profile, winter ratings (50F rating), interchange, and accompanying generation dispatch. The study area of the model is adjusted, and the appropriate winter model is used for neighboring system areas. Interchange levels for the various system zones reflect all yearly *long-term firm* (LTF) transmission service, or the historical average. Load level, interchange, and generation dispatch for external areas impacting system facilities are based on statistical averages for previous winter peak periods. Thus, the same network model and criteria used for the reference system case apply. The flowgates ultimately used in the winter peak reliability analysis are determined by running all applicable contingencies and monitoring all market monitored facilities and all BES facilities. The same single contingency power flow solution techniques used in other reference system reliability tests also apply.

10.3.3.9 Spare Equipment Availability Outages of certain major transmission equipment require the long maintenance period which requires a replacement by a spare one as soon as possible. Sometimes, the replacement of an equipment, such as power transformer, can have a lead time of more than 1 year. Therefore, it is necessary to review and evaluate the unavailability of such an equipment and assess the impact of this possible unavailability on system performance using applicable contingencies. This assessment should consider the conditions that the system is expected to experience during the possible unavailability of the long lead time equipment. It may be necessary to conduct the economic impact of unavailability of such major equipment. The results from this economic analysis can help determine whether it is economic to purchase and maintain a spare equipment to maintain system reliability.

10.3.3.10 Maximum Credible Disturbance Review   Extreme events are sometimes inevitable. Such events known as maximum credible disturbances, should be reviewed and their impact on the system reliability must be assessed. If the initial analysis shows cascading caused by the occurrence of extreme events, the system planner should perform an evaluation of possible actions designed to reduce the likelihood or mitigate the consequences and adverse impacts of the events. This can include a stability analysis of the area. Therefore, the impact of extreme events using stability analysis must be assessed.

10.3.4 Long-Term Reliability Review

In addition to near-term reliability analysis, long-term reliability review is necessary. The longer term planning horizon can span the current year plus 6 through the current year plus 15 or more. Reference cases for some or any future years can be developed and evaluated. Assumptions and model development regarding this longer-term view have to be developed.

The longer-term view of system reliability is subject to increased uncertainty due to the increased likelihood of changes in the analysis as time progresses. The purpose of the long-term review is to anticipate system trends which may require longer lead time solutions. This enables an ISP to take appropriate actions when system issues may require initiation during the near-term horizon in anticipation of potential violations in the longer term. System issues uncovered that are amenable to shorter lead time remedies will be addressed as they enter into the near-term horizon.

In the 15-year out analysis, the longer term reliability review involving single and multiple contingency analyses is conducted to detect system conditions which may need a solution with a lead time to operation exceeding 5 years. Two processes can be used as indicators to determine the need for contingency analysis in the longer term horizon. The first is a review of the near-term results to detect violations that occur for multiple deliverability areas or multiple or severe violations clustered in a one area of the system. This review may suggest larger projects to collectively address groups of violations. The second is a thermal analysis including double circuit tower line outages at voltages exceeding 100 kV performed on the current year plus 15 system.

All of the current year plus 15 results produced will be reviewed to determine if any issues may require longer lead time solutions. Those solutions can be identified.

In this evaluation of the need for longer lead time solutions, load shedding and/or curtailment of firm transactions may be employed to ease potential violations. Due to uncertainty associated with longer time horizon, linear DC analysis is more appropriate. The analysis for the longer-term horizon evaluates all single contingencies against the same normal and emergency thermal ratings criteria used for the near term. Reactive power analysis may also be done as part of the long-term analysis.

10.4 MARKET-BASED TRANSMISSION PLANNING

Generally, in a market environment, transmission-based solutions produced by transmission planning, and generation-based solutions, such as merchant generation, are not just competing, but also complementing each other. Hence, transmission planning has become even more important and complex in the competitive market environment. In this environment, the traditional approach using integrated resource planning (IRP) is no longer applicable. There are also higher levels of uncertainty for a long-range transmission planning. For these reasons, transmission planning must include not only traditional reliability analysis but also economic analysis under a wide range of possible future scenarios on a probabilistic basis.

Similar to the approach used in the conventional transmission planning, market-based transmission planning includes both reliability and economic analysis. Reliability analysis, which focuses on the technical assessment of system reliability, does not change much from the conventional approach. However, the economic analysis part has changed significantly because of the nature of the operating markets especially for those which use the nodal-based pricing for congestion. One of the primary purposes of conducting economic analysis is to estimate projected economic benefits that proposed transmission projects can bring into the system. Then, this estimated benefit is compared with the estimated cost of the project. The transmission projects whose economic benefits far exceed their costs provide the best economic value to the system. These projects are chosen to become market-based transmission solutions.

10.4.1 Market Efficiency Model

Transmission planning can be done not just for maintaining reliability but also for reducing congestion in the electricity market regime. Note that a robust transmission system can enhance competition and mitigate market power in a restructured market environment. An economic model of market-based transmission planning is also known as the *market efficiency model*. Market efficiency analysis studies entail production cost analysis to evaluate proposed transmission enhancements for their economic value based on their ability to relieve persistent transmission congestion.

Reduction in congestion in a market system will increase the consumer surplus and thus increase the social welfare.

The salient feature of this model is that it is based on the forecasted congestion in the market system for future years, based on the best possible system topology and generation scenario, while meeting the current reliability criteria with minimum reliable reserve margin. Required generations for future years are added into the system based on the current pattern of market-based generation expansion. The benefits and costs of proposed transmission upgrade projects are compared by using reasonable estimate of discount rates and the net present value (NPV) method to determine the economic viability of the transmission upgrades. Sensitivity analyses are typically done to determine the various key drivers for influencing the economic outcome. Market efficiency analyses have to be done according to the established benefit-to-cost metrics.

The new regime under the electricity market, which uses the scheme of efficient and real-time pricing of transmission via nodal pricing (LMP pricing), has provided new opportunity based on which economic criteria can be developed to help choose a market-based transmission upgrade which is more economic, in addition to meeting the current and future reliability criteria. Economic analysis, using projections from locational based market prices, will show the overall economic benefit of the system reinforcements in terms of changes in energy prices, revenue to individual generators, payment by load customers, and total system production cost.

Economic analysis under the market-based model requires the simulation of the power system operation which occurs in a competitive market environment. In other words, the simulation of the power system should mimic the actual market operation, using similar unit commitment and economic dispatch algorithms. To that end, the market simulation part of economic analysis uses the security-constrained unit commitment (SCUC) and security-constrained economic dispatch (SCED) algorithm, which is at the core of market operation, determination of LMP, and congestion pricing. Readers can refer to previous chapters for the detailed background of this algorithm. To evaluate the economic impact of a transmission upgrade, simulations of power market operation are done *with* and *without* that upgrade. The process diagram including the input data necessary for doing this type of market simulation is shown in Figure 8.5 in Chapter 8.

10.4.2 Model Benchmarking

One of the key steps required in conducting market simulation for economic analysis is to benchmark the model that is being built. The analyst engineer or modeler should be cognizant of the fact that it is always a challenge to model a future system. Therefore, it certainly helps if the model is tested and benched against known results from the historical years. To effectively achieve this goal given the complexity of the model, the market results from the previous historical year are typically used in conducting model benchmarking because there were already known inputs and output from the same system. From historical market and system operation, the following information will be useful for model benchmarking:

1. Generation output by each generator
2. Congestion on each binding transmission constraint
3. Market prices (day-ahead, real-time)
4. Transmission outages
5. Generation outages
6. Other key system parameters

Using all the known input parameters, the model can be simulated for a historical year to see if the model produces reasonable results for key system output, such as generation output by generator types, average market prices, binding transmission constraints, and total amount of congestion values. Benchmarking historical tie line flows will also be important if a model includes more than one pool or market system. By this way, reasonable MW amount of cross-border power transfer can be properly captured in the model. Recall that the hurdle rates among the pools are used to produce reasonably matching cross-border flows from the model. It would be helpful if the modeler is familiar with the specific institutional setup or rules of the pools or system that is being modeled.

10.4.3 Market Efficiency Analysis

The flowchart for conducting an economic analysis of proposed transmission projects is shown in Figure 10.3. The key steps involved in conducting an economic analysis of transmission projects are outlined below.

1. Identify congested transmission constraints
2. Propose transmission upgrades that can potentially relieve the identified congestion
3. Conduct market simulation with and without the proposed transmission upgrades for either all years or selected years of the transmission project life
4. Estimate projected economic benefits for simulated years and convert these benefits (in current years) into their equivalent present values
5. Estimate the projected cost of the proposed transmission upgrade and convert it into its equivalent present value
6. Compute the benefit-to-cost ratio of the proposed transmission upgrade and compare it with the established metrics

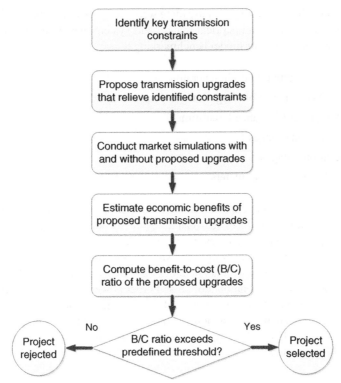

Figure 10.3 Flowchart for Conducting Economic analysis of Proposed Transmission Projects.

  7. Select the proposed transmission upgrade if its benefit-to-cost ratio exceeds the predefined threshold

In Step-3, ideally, simulations must be done for every single year that the proposed upgrade will be in service. However, due to the computational burden and the complexity of input data preparation, actual market simulations can be done just for a few representative years. In this case, results for intermediate years can be obtained by interpolating the results from years for which explicit simulations are done. Note also that the simulation results for the years that are far into the future tend to exhibit higher level of uncertainty.

After market simulation is done for each representative year in Step-3, the generation schedules and LMP prices for each node in the system can be obtained. LMP is composed of three components: system energy price, congestion, and loss. LMP prices are used to compute congestion, revenue to each generator, and payment by load.

10.4.4 System Economic Output

Before the possible ultimate economic criteria used in the economic analysis are defined, it is important to define some relevant system economic variables.

10.4.4.1 Production Cost The production costs are the costs to generate energy at the desired level of output for each simulation period to meet demand. The generator production cost is the summation of the hourly fuel cost, operation and maintenance cost, startup cost, and emission cost for each generating unit in the system. Generally, the fuel cost is the largest component of total production cost. All variables in the following equation have the unit of $/MWh.

$$PC_i = FC_i + OM_i + ST_i + EC_i \qquad i \forall G, \tag{10.1}$$

where PC_i is the production cost of generator i, FC_i is the fuel cost of generator i, OM_i is the operation and maintenance cost of generator i, ST_i is the startup cost of generator i, EC_i is the emission cost of generator i, and G is the total number of generators.

10.4.4.2 Load Payment The load payment is defined as the product of the hourly energy (MWh) consumed at each load location and the hourly LMP ($/MWh) at that location summed over the simulation period.

$$LP_k = L_k \times LMP_k \qquad k \forall D, \tag{10.2}$$

where LP_k is the load payment at bus k ($), L_k is the load demand at bus k (MWh), LMP_k is the hourly LMP at bus k ($/MWh), D is the total number of demand nodes.

10.4.4.3 Generator Revenue The generator revenue is defined as the product of the hourly energy (MWh) output for each generating unit and the hourly LMP ($/MWh) at the generator's location summed over the simulation period. Generator profits can be calculated by subtracting the generator production costs from the generator revenue.

$$GR_i = G_i \times LMP_i \qquad i \forall G, \tag{10.3}$$

where GR_i is the generator revenue at generator bus i ($), G_i is the energy generation at generator bus i (MWh), LMP_i is the hourly LMP at generator bus i ($/MWh), G is the total number of generators.

10.4.4.4 Congestion Cost Congestion cost between two nodes is the difference between LMP prices of these two nodes, ignoring marginal loss component.

Congestion cost is typically incurred by load customers, while generators receive congestion credits.

$$CC_{ik} = LMP_i - LMP_k, \tag{10.4}$$

where CC_{ik} congestion cost between bus k and bus i ($/MWh), LMP_i is the LMP at bus i, and LMP_k is the LMP at bus k.

10.4.5 Various Definitions of Economic Benefits

Using the fundamental system outputs from the simulation, several metrics for economic benefits of transmission upgrades can be defined. One should note that the development of and the decision to use these metrics are primarily based on the decision-making of a particular electricity market and the input from its stakeholders. The following definitions can be defined for estimating the economic benefits of transmission upgrades under the competitive market environment:

1. Saving in production cost only
2. Composite saving of both production cost and load payment, with some weights
3. Composite saving of production cost, load payment, and congestion cost, with some weights
4. Any combination of any key system output variables

For example, the following composite measure of saving (reduction) in two major variables can be used as an economic benefit:

$$\text{Economic benefit} = \alpha \times (\Delta PC) + \beta \times (\Delta LP), \tag{10.5}$$

where $(\alpha + \beta) = 1$; α, β are weighting factors, ΔPC is the change (reduction) in production cost due to a transmission upgrade, ΔLP is the change (reduction) in load payment due to a transmission upgrade.

10.4.6 Economic Criteria

In any economic analysis, benefits must exceed the costs to add economic value to the system of interest. In the economic analysis of evaluating transmission projects, the estimated economic benefits of a proposed project must exceed its cost. To make meaningful comparison between economic benefits and cost, the annual economic benefit which is the composite saving for a year must be converted to a present value using appropriate discount rate. The annual carrying charge of the transmission upgrade under consideration, which is the annual cost of the project, should also be converted to a present value using the same discount rate. This method is known as net present value (NPV) method. Typically, benefit-to-cost ratio of 1.25 is used as a threshold for screening proposed transmission projects, where 25% is accounted for future uncertainty. In other words, transmission projects whose benefit-to-cost ratio (in present value) exceeds this threshold are selected as market-based transmission solutions.

10.5 CROSS-BORDER TRANSMISSION PLANNING

Sometimes, it is necessary for a power system to interconnect or expand transmission ties with a neighboring system. A transmission interconnection or expansion plan that crosses the borders of two or more adjacent systems is called cross-border transmission planning. Transmission planning within the closed boundary of a system is challenging enough. And yet, transmission planning that crosses the border between two adjacent systems would be even more challenging. Cross-border transmission planning at the border of two or more adjacent systems is done for at least two major reasons: (1) reliability and (2) economics. Interconnected system can improve the reliability of both systems. For example, if one system is facing an emergency situation such as shortage of power supply, then the neighboring system can provide extra supply via transmission ties. Economic reasons include facilitation of economic power transfer between the two systems and reduction of cross-border congestion between two market systems. Other reasons for cross-border transmission planning include dedicating a transmission interconnection between the two systems to transmit the supply power from one system to serve a specific load in the adjacent system. For example, there are several DC ties from neighboring systems into the load center in NY City. Transmission ties across the system boundaries or national borders are fairly common around the world. However, such a cross-border transmission planning can take a long time to be implemented because each party (system) has to agree on the terms and conditions of interconnection plan, interconnection rules, investment, and joint operating rules. Cross-border transmission planning poses a unique challenge for two neighboring systems with different jurisdictional requirements.

10.5.1 Reliability

The principal objective of reliability-based transmission planning for a system is to ensure that there is a sufficient level of transmission investment in the system that is able to support the transmission of power from generation resources and the delivery of power to locations where power is consumed. In doing so, the system should also be able to withstand a loss of at least one facility, such as a transmission line, a transformer or a generator—called $(N-1)$ criteria—without having any adverse consequences related to reliability issues. Transmission planners also have to continuously assess the system condition targeted for a future period given all possible uncertainties, such as uncertain generation from renewable energy sources, uncertain generator retirement, and uncertainly growing load. Occasionally, interconnecting the two adjacent power systems by transmission ties can improve the reliability, robustness, and resiliency of the larger system. This interconnection tie would allow some emergency power transfers from one system to another in case one system is experiencing a generation shortage.

10.5.2 Economics

Another reason for implementing transmission ties across the border is based on the economic benefit. For example, assume that one system has excess amount of lower cost generation supply and its neighboring system has higher cost generation supply. In this case, both systems have an incentive to interconnect or expand the ties to facilitate the economic power transfer from one system to another. The power transfer is generally from the lower cost system to the higher cost system.

Related to this economic power transfer is the concept of congestion reduction across the two adjacent market systems. For example, there can be transmission congestions that appear at the border of two adjacent market systems (such as between PJM and MISO). Even though some kind of monetary exchange approach can be implemented on a daily basis to settle for the cross-border congestion, congestion for a subset of the cross-border constraints continues to exist. We can call this *chronic cross-border congestion*. In this particular chronically congested situation, a more efficient solution appears to be building a transmission tie or interconnection across the two systems. As congestion represents a reduction of consumer surplus, the ISPs have an incentive to build transmissions that would relieve that congestion so as to reduce the associated congestion cost. The primary purpose of this kind of transmission planning is to increase the social welfare by increasing the market efficiency. Reduction of congestion by adding transmissions means reduction of load payment by consumers, which would increase consumer surplus.

However, since this type of transmission planning involves two or more different planning authorities or system planners, a common methodology that is applicable and acceptable to both or all parties needs to be developed. Such a methodology for promoting cross-border transmissions at the seams of two adjacent electricity market systems is subsequently presented.

10.5.3 Methodology for Cross-Border Transmission Planning

A cross-border transmission can be any transmission or transmission-related project that will be planned and likely to be built at the intersection of any two or more transmission systems or market systems. Such a cross-border transmission facility should *physically* cross the border of two or more adjacent and contiguous electricity markets. It must also be noted that this is not a single market problem as the two electricity markets are adjacent and distinct. This is the unique nature of the problem.

As mentioned earlier, there are many primary and secondary benefits associated with cross-border transmission projects. For example, the primary benefits include relieving cross-border congestions which increases the social welfare. The secondary benefits include facilitating the transfer of more economic power from one area to the next, and reinforcing reliability for one or more transmission systems and thus improving the security of power system operations. The method is based on the economic criteria that jointly apply to two neighboring electricity markets across which the transmission projects will be proposed, planned, and built. The method is also different compared with the current methodologies for planning either market-based or reliability-based transmission projects within each market system. It is a workable methodology that would support the plan for cross-border transmission that would help both market systems in achieving the dual goal of enhancing system reliability and improving economic efficiency. This method is also applicable to other electricity markets or transmission regimes where such cross-border transmission expansions are legitimately needed.

This methodology is based on a number of assumptions related to two adjacent systems which are stated below:

1. There must be functioning electricity market in each of the two systems.
2. These markets should operate both day-ahead and real-time markets based on a nodal pricing scheme.
3. The markets may or may not have mandatory capacity markets, financial transmission rights (FTR) market, and other ancillary service markets.
4. At least, one market system has a workable method on the economic analysis of transmission projects proposed within its own system.
5. Each market system must have an ability and authority to approve a cross-border transmission project that passes the pre-determined economic criteria, so that this project will be built eventually.

Figure 10.4 shows the methodology framework for economic analysis of cross-border transmission planning in the context of organized electricity markets. From the figure, *Base Case* represents the current or realistic future transmission system for combined market systems. Using information on network, generation, demand, and constraints as input, market-clearing optimization is done to produce various economic outputs. *Change Case* represents a new transmission system with an addition of a new cross-border transmission project under consideration. For the *Change Case*, the same market-clearing optimization is done using the new network which includes the proposed cross-border transmission project and a new set of constraints.

The key input data and steps involved in this framework are further explained.

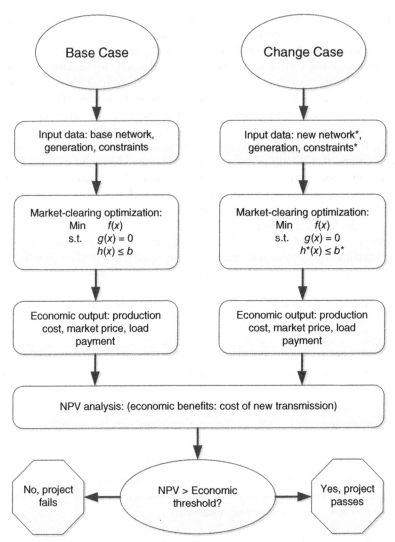

Figure 10.4   Framework for Economic Analysis of Cross-Border Transmission Planning.

10.5.3.1 Network Topology Modeling   Network topology models should
be represented by the best available model for each of the two systems of interest.
The network model for *Base Case* should represent the combined network model of
two systems without the proposed cross-border transmission project. The network
model for *Change Case* represents the combined network model of two systems with
the proposed cross-border transmission project.

10.5.3.2 Generation Data Modeling

Resource mix should be represented by the addition and retirements of generators of both systems. The installed generation must meet the planned reserve margin requirement so as to avoid shortages and emergency generation. The required generation for future years can be added proportional to queue units based on location and fuel type, which are driven by market signals. Other methods for adding future generations can be used as applicable. These added generation units with existing units make up the entire generation portfolio in the model.

10.5.3.3 Constraints Modeling

Modeling of transmission constraints is critical to any market-clearing optimization framework. In addition to modeling credible transmission constraints from each electricity market system, it is also important to model the set of constraints that typically appear right at or near the border of the two markets. For example, this set of constraints between PJM and MISO markets is called "market-to-market (M2M) constraints." For example, this set of M2M constraints also includes transmission constraints that are located within each market system but near the border. These constraints are included in M2M set because generation from each market system can have an impact, sometimes significantly, on the binding of these constraints. In selecting this set of M2M constraints, a screening method based on historical market settlement dollars (special arrangement between PJM and MISO) or a screening method based on historical congestion dollars or combination of the two methods can be used as screening criteria. Once selected, this set of constraints combined with other constraints form the full set of constraints to be used in market-clearing optimization.

10.5.3.4 Market-Clearing Optimization

The objective function $\{f(x)\}$ of market-clearing problem is to minimize the bid-based production cost of the combined market system, subject to equality constraint $\{g(x) = 0\}$ and inequality constraints $\{h(x) \leq b\}$. The equality constraint represents the balance of generation and load to ensure a reliable operation of power system. The inequality constraints explicitly model the transmission network constraints and generators' operational constraints. The goal is to ensure that the simulated flows on the network are within the specified limits and generator outputs are also within their operational limits. The key output of such market-clearing optimization includes production cost for each system or a combined system, market prices for each electrical node in each or combined system, congestion (in both dollars and hours), and other derived output such as generator revenue and load payment.

10.5.3.5 Economic Criteria

Based on these system outputs from the market-clearing optimization, the next step is to estimate the economic benefit of adding a cross-border transmission line. Each of the possible economic benefit measures from such a simulation is presented in Section 10.4 Those same measures can be used in estimating economic benefits of adding a cross-border transmission for two adjacent systems. Once economic benefits of adding a tie line are estimated, the ratio of benefit-to-cost of the project is further computed to see if that project meets or exceeds the required threshold, using appropriate discount rates and NPV analysis. This kind of economic criteria can be developed for the unique purpose of selecting cross-border transmission plans. Other possible approach may include a method which would require the proposed transmission project to pass both joint economic criteria and individual economic criteria of each respective market system.

While the method is presented in the context of nodal electricity markets in the United States, the method is also applicable to other electricity markets, such as a European electricity market. However, it may be necessary to do some modifications in terms of assumptions and modeling to make this method work for such a market. In fact, this method is generally applicable to any other transmission regimes with electricity markets where such cross-border transmission expansions are desperately needed.

In addition, it is possible to develop an alternative method in which an optimization problem can be formulated and mixed integer programming algorithm (MILP) can be used to find an optimal solution for two-markets problem, treated as a single market problem. The optimal solution will be the solution to this unique problem of cross-border transmission planning.

10.6 COMPETITIVE TRANSMISSION PLANNING

Exposing the transmission planning under an organized market regime into a competition is a fairly new development. This is primarily driven by a recent order from the US federal regulatory authority. Even though this competitive process for transmission planning is not mandatory, some ISPs are already developing competitive bidding processes to solicit competitive transmission solutions from any eligible designated entity. This process will create plenty of opportunities for nonincumbent transmission developers to compete with incumbent transmission owners for the right to build, own, and operate a transmission facility. Consequently, more innovative transmission solutions can be developed in the near future.

10.6.1 Regulated Transmission Business

Transmission grid is a public good because every electricity consumer connected to the grid benefits from it. By analogy, the transmission grid plays a similar role as highway roads and bridges as part of a nation's infrastructure. Up until now, transmission business is treated as a regulated business because it is a natural monopoly. In a natural monopoly, only one efficient firm is possible. In other words, it would be redundant and grossly inefficient if two different transmission companies attempt to serve the same load by building two parallel networks. Therefore, it makes economic sense if only one set of transmission network is used to serve load for a particular defined area. That transmission network may be owned by a single utility company. Transmission network is also an infrastructure investment based on regional or utility planning with cost recovery at regulated rates.

The entire transmission business from project proposal, construction, financial responsibility, and ownership, to operation and control of a transmission facility in a traditional utility service area is done by the regulated incumbent utility (natural monopoly). In the restructured market environment, the operational control of all bulk-power transmission facilities falls into the authority of an ISO. However, the incumbent transmission utilities continue to be responsible for project proposal, construction, financial responsibility, and ownership. These utilities are also responsible for proper maintenance of their transmission facilities.

For transmission planning in a non-organized market area, the regulated incumbent utility is solely responsible for planning the transmission system in its own operating territory. However, for transmission planning in an organized market area, the ISP, in joint coordination with incumbent transmission-owning utilities, is responsible for planning for the transmission system. The major steps in this transmission planning process are to

1. Identify the issues and drivers for transmission needs
2. Propose multiple transmission solutions that will potentially resolve these issues
3. Conduct reliability, economic, and other analyses as necessary
4. Select and approve the final transmission solutions
5. Construct the approved transmission projects

While the process is transparent and open for stakeholders' feedback, the regulated incumbent transmission-owning utilities generally received the right to build such final approved transmission projects. The opportunity for the nonincumbent transmission developers to receive the right to build a new transmission line in a market environment is quite slim.

10.6.2 FERC Order 1000

To remedy this and other issues, US federal regulatory authority issued an order, Order 1000, in July 21, 2011. While FERC order requirements include other key issues, such as cost allocation and interregional reforms, our focus in this section is on how this order can pave the way for nonincumbent transmission developers to participate in the competitive process for planning transmission, especially in an ISP-operated region.

In FERC Order 1000, there are specific requirements related to nonincumbent transmission developer. Three key rules related to nonincumbent transmission developer in this order are as follows:

1. Promotes competition in regional transmission planning processes to support efficient and cost effective transmission development

2. Requires the development of a not-unduly discriminatory regional process for transmission project submission, evaluation, and selection

3. Removes any federal right of first refusal from Commission-approved tariffs and agreements with respect to new transmission facilities selected in a regional transmission plan for the purposes of cost allocation, subject to four limitations: (a) this does not apply to a transmission facility that is not selected in a regional transmission plan for purposes of cost allocation; (b) this does not apply to upgrades to transmission facilities, such as tower upgrade/replacement or reconductoring; (c) this allows, but does not require, the use of competitive bidding to solicit transmission projects or project developers; (d) Nothing in this requirement affects state or local laws or regulations regarding the construction of transmission facilities, including but not limited to authority over siting or permitting of transmission facilities.

FERC Order 1000 essentially challenged the monopolistic nature of transmission business in the industry by attempting to remove this right of first refusal with respect to new transmission facilities selected in a regional transmission plan for the purposes of cost allocation. Order 1000 envisions a level playing field on which new transmission developers can compete with established transmission utilities for the right to build new transmission lines. The Commission was concerned that "federal rights of first refusal create opportunities for undue discrimination and preferential treatment against nonincumbent transmission developers within existing regional transmission planning processes." FERC also believed "that expanding the universe of transmission developers offering potential solutions can lead to the identification and evaluation of potential solutions to regional needs that are more efficient or cost-effective." The only exceptions were to be for local projects that are not subject to regional planning or cost allocation, upgrades of existing transmission facilities, and currently planned transmission projects. FERC has given regional transmission organizations the flexibility to solicit sponsors for transmission upgrades through competitive bidding procedures, but did not mandate this approach.

10.6.3 Competitive Bidding Process

Although FERC Order 1000 does not require the use of competitive bidding process to solicit transmission projects in an ISP-operated region, different ISOs are at different stages of implementing various forms of competitive processes, largely

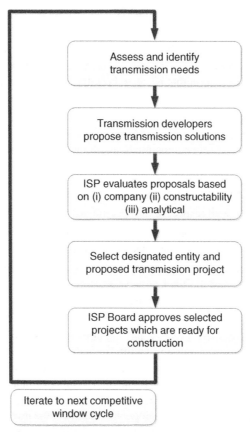

Figure 10.5   Flowchart for a New Competitive Process for Soliciting Transmission Projects.

as a result of this order. For example, PJM as an ISP, has started to develop a process, which can be called *competitive transmission planning*. This new competitive process provides opportunity for nonincumbent transmission developers, on an equal footing with incumbent transmission companies, to submit project proposals through a *proposal window* and be considered for project construction, ownership, operation, and financial responsibility. During each window, the ISP seeks transmission proposals to address one or more identified needs: reliability, market efficiency, operational performance, and public policy, for example. Once a window closes, the ISP proceeds with a specific company, analytical and constructability evaluations to assess the proposals submitted and recommends a solution to the Board. Figure 10.5 shows a flowchart for this new competitive process in soliciting transmission projects from both incumbent and nonincumbent transmission developers.

The key steps of this process are outlined below:

1. ISP opens a window (from 30 to 120 days) within which transmission developers (both incumbent and nonincumbent) prepare and submit project packages. The project information that needs to be submitted include a detailed project plan, cost of the project, and timeline to complete the project.

2. Once the proposal window closes, the ISP conducts three kinds of evaluations: (a) company evaluation which assesses a designated entity's ability to construct, own, operate, maintain, and finance a specific project; (b) constructability evaluation which assesses proposals in terms of cost, schedule, siting, permitting, rights of way, land acquisition, project complexity, and coordination risks; (c) analytical evaluation which assesses proposals in terms of performance with respect to the specific, identified needs.

3. Throughout the evaluation process, the stakeholders are apprised of the status of the evaluations and their feedbacks are sought.

4. After the evaluations are complete, the final designated entity (the transmission developer) and the final projects are selected. The ISP board approves the selected developer and projects.

With this new competitive process, any incumbent transmission company can build and own a piece of transmission facility in the territory of another incumbent transmission company. The nonincumbent transmission developer will also have that same opportunity. Due to this new development, a number of independent transmission developing companies emerge in the United States. In contrast to the merchant transmission projects for which the developer has to pay for the investment, the investment cost for the new proposed projects in this competitive process will be borne by regulated rates, once approved. Therefore, the return on investment for such transmission projects is guaranteed as any other existing regulated transmission business. This guaranteed rate of return on regulated transmission investments will certainly spur more activities by both incumbent and nonincumbent transmission developers. At the same time, more innovative, cost-effective, and efficient transmission solutions can be developed.

10.7 TRANSMISSION PLANNING IN EUROPE

All *transmission system operators* (TSOs) in Europe share some concerns related to the optimization of the internal energy market, climate change, and the high penetration of renewable energies in European energy system. In addition, system operators wanted to develop a more demand-centric approach and more flexible energy interchange, improving the competitiveness of the energy markets. In 2009, under regulation 714/2009, the *European Network Transmission System Operator for Electricity* (ENTSO-E) was established. It represents 42 TSOs from 35 different European countries. The initial goals of this organization were

1. **Sustainability**: Recently, the amount of renewable energy resources has increased considerably in the entire Europe including both ends of the continent. Large amount of power flows across all Europe. Therefore, it is very important to facilitate the integration of these resources throughout the European transmission system.

2. **Market Integration**: Since European transmission network comprises several systems, it is challenging to integrate the markets that have different market rules.

3. **Market Competitiveness**: Market competitiveness allows the market decentralization with respect to market participants, setting the right incentives to motivate them to participate, and at the same time reduce their impact on sensible issues such as flexibility, capacity reduction or price volatility.

4. **Reliability**: The forecasting tools should be improved in order to produce a reliable calculation of both demand and renewable resources energy output.

5. **Emission Reduction**: Significant reduction of CO_2 or simply carbon reduction is set as one of the main objectives.

From the initial objectives, the main tasks were

1. Management and coordination of the European electricity transmission network

2. Network Code: Establishes a set of rules that applies to different areas of the energy sector. These areas are generators, congestion management, demand, operation security, load frequency, electricity balancing, and high-voltage direct current connections.

3. Common Information Model (CIM): In order to interchange network data among all participants, it is important for all TSOs to generate consistent and coordinated data. That is the reason they adopted the IEC CIM model to interchange data for studies. Interoperability tests are performed every year to validate the interoperability of its CIM standard and the IEC standards. In addition, ENTSO-E supports CIM development for both grid models and market exchanges.

Nowadays the development effort is focused in four main areas: (1) storage devices integration, (2) new monitoring techniques, (3) better control systems, and (4) integration of demand response.

10.8 FURTHER DISCUSSIONS

There are a number of emerging issues that are going to impact the current practice of transmission planning process.

At least in North America, the weather patterns have been more and more severe and extreme even with low probability of occurrence. Those severe weather events such as superstorms, hurricanes, tornadoes, and extremely cold weather, can cause significant damages to the energy infrastructure including power grid. Possible aftermaths include generator outages, power supply interruption, and substation flooding, all leading towards partial blackouts. Therefore, it is necessary for system planners to assess their power systems to identify the weakest areas of the grid and to plan to strengthen the grid. The objective is to make the power grid more resilient so as to withstand the impact from such extreme events in the future. Some electric utilities, severely hit by those events, are already planning to spend from millions to billions of dollars to harden their power system particularly the transmission and distribution systems.

Recent events of extremely cold weather have shed light on the interdependence between natural gas delivery and electric supply for systems that have relied heavily on gas-fired units. The failure of one system can significantly impact the other. Interruption in the gas supply system can make a number of gas-fired units unavailable, which can cause power shortages, price spikes, and threaten the system reliability. System planners have to take a hard look at planning for a resilient transmission system which can withstand such power supply interruptions. One area for improvement is to model contingency events in conducting transmission planning associated with a situation in which the outage of natural gas supply system occurs for any reason.

Another emerging development is that a massive amount of renewable energy sources is going to be connected to the power grid in the near future. This trend will force transmission system planners to seriously evaluate and reevaluate the options to accommodate the growing penetration of RES. Effective approaches should be developed to deal with this emerging issue.

Until recently, the magnitude of uncertainty in the transmission-planning process regarding future system conditions was limited. Tests used in the transmission planning process could reasonably define the expected date of future reliability violations. System planners could plan new transmission facilities with minimal risk of fluctuating dates marking the expected onset of reliability standard violations. That has changed in many respects. A single set of baseline and market assumptions are simply not sufficiently flexible to consider all emerging operational, economic, and regulatory trends. The increase in the level of uncertainties associated with various analysis assumptions would require system planners to conduct more and more scenario-type of analyses to better forecast the system trend in the future. Based on the outcomes of these scenarios, the transmission system has to be planned accordingly. It will certainly be a challenge to manage these emerging, sometimes diverging, trends.

CHAPTER END PROBLEMS

10.1 Today's date is October 2016. During the routine system reliability analysis, a low-voltage situation was found at the 230 kV substation if a nearby 345 kV transmission line is removed from service for the period of 2018 summer. You plan to install some capacitor banks at the 230 kV substation to meet the voltage guidelines of your planning criteria. Will that upgrade constitute a baseline reliability project?

10.2 Today's date is October 2016. A generator merchant company is proposing a new generator at a 138 kV substation to be installed in January 2020. You do your routine generator interconnection study and find that the connection of this new generator will cause a reliability violation. To prevent this violation, you propose a transmission upgrade near the connection point. Will that upgrade be considered as a baseline reliability project?

10.3 Find out a set of voltage criteria from a specific planning authority. Why are the required voltage limits different for some voltage levels?

10.4 Competitive transmission planning process was designed to provide a level playing field for an opportunity to propose, build, own, and operate a new transmission facility for both incumbent transmission-owning companies and nonincumbent transmission developers. Despite this effort, the incumbent transmission companies have several competitive advantages over nonincumbent developers. Can you identify at least three such advantages? Explain why.

10.5 In an electricity market, a chronic congestion occurs between two adjacent nodes A and B. The independent system operator (ISP) responsible for transmission planning for this market area is evaluating three proposed transmission projects that can relieve this particular congestion. The ISP ran a number of market simulations and came up with the system benefits for the first 10 years of the projects shown in Table 10.1 (For simplicity, assume the benefits/costs are considered only for 10 years.)

Assume discount rate of 7%; fixed carrying charge rate of 15%; Project X cost = $20 million; Project Y cost = $30 million; Project Z cost = $15 million.

a. Rank the proposed projects in the order of preference based on the net present value of benefit-to-cost ratio.

b. Which project should the ISP select if the required threshold of benefit-to-cost ratio is 1.25?

c. If more than one proposed projects exceed the minimum required economic criteria, how should the final single project be selected? and why?

d. Redo subproblems (a), (b), and (c) if the discount rate is assumed to be 5%.

Table 10.1   System Benefits (million $) of Each Proposed Project for First 10 Years

Projects	Yr-1	Yr-2	Yr-3	Yr-4	Yr-5	Yr-6	Yr-7	Yr-8	Yr-9	Yr-10
Project X	2.8	2.9	2.9	2.4	2.6	2.5	2.5	3.0	2.6	2.8
Project Y	6.3	6.0	6.8	5.8	5.7	6.1	5.6	6.6	6.7	6.4
Project Z	3.0	2.8	2.6	2.7	2.7	2.8	3.2	3.1	3.2	3.0

FURTHER READING

1. Li W. *Probabilistic Transmission System Planning*. Hoboken, NJ: Wiley-IEEE Press; 2011.

2. Chow JH, Wu FF, Momoh JA, editors. *Applied Mathematics for Restructured Electric Power Systems: Optimization, Control, and Computational Intelligence*. Springer; 2005.

3. *Manual 14B: PJM Regional Transmission Planning Process*. PJM Interconnection. Available online at http://pjm.com/

4. Lin J. Market-based transmission planning model in PJM electricity market. In: Proceedings of the 6th International Conference on the European Energy Market (EEM09), May 2009, pp. 1–6.

5. Lin J. Modeling and simulation of PJM and Northeastern RTOs for interregional planning. In: Proceedings of IEEE Power Engineering Society General Meeting, July 2013, pp. 1–6.

6. Lin J. Cross-border transmission planning in two adjacent electricity market systems. In: Proceedings of the 12th International Conference on the European Energy Market (EEM15), May 2015, pp. 1–6.

7. *Transmission planning and cost allocation by transmission owning and operating public utilities, Order No. 1000*. U.S. Federal Energy Regulatory Commission Final Rule, 2011.

8. The European Network of Transmission System Operators, ENTSO-E. [Online]. https://www.entsoe.eu/

Chapter 11

Electricity Market under a Future Grid

Power system is evolving. Some of the recent changes in the power system are quite dramatic. Growing penetration of renewable energy technologies, development of smart grid including microgrid, electric vehicles, advanced energy storage system and smart sensors, growing number of *distributed energy resources* (DERs), such as smaller-scale generators, microturbines, and fuel cells, connected to the distribution system, are some of the emerging developments in the power grid. Some electricity markets are already dealing with, or are in the process of dealing with such emerging issues. In this chapter, we will describe these emerging issues and their potential impact on current electricity markets. Obviously, these emerging developments are active areas of research. Opportunistically, we will also make some speculations about the future grid.

11.1 RENEWABLE ENERGY TECHNOLOGIES

Triggered by the greenhouse gas (GHG) emissions, the threat of climate change has galvanized many nations to look for a cleaner source of energy. By combination of public and private endeavors, the costs of clean energy technologies have recently fallen, which led to a surge in demand and deployment of such technologies. Most of the clean energy technologies are renewable energy technologies which represent any technologies that generate electricity based on renewable energy sources (RES). According to the US Energy Information Administration (EIA)'s latest reports, the electrical generation from non-hydro renewable energy sources (i.e., wind, solar, biomass, geothermal) has expanded in recent years. Based on their short-term energy forecast, electricity generated from utility-scale renewable plants is expected to grow

Electricity Markets: Theories and Applications, First Edition. Jeremy Lin and Fernando H. Magnago.
© 2017 by The Institute of Electrical and Electronics Engineers, Inc. Published 2017 by John Wiley & Sons, Inc.

to 9% in 2017 while that growth will reach nearly 10% in 2018. Much of the growth comes from new installations of wind and solar plants and increases in hydroelectric generation. Increases in renewable resource capacity and generation are influenced by federal, state, and local policies. According to a report by International Energy Agency (IEA), electricity production by renewable energy sources is expected to grow in many regions of the world.

11.1.1 Types of Renewable Energy Sources

Generally, RES include solar electric, solar thermal, wind, hydroelectric power, geothermal, ocean current, or wave energy resource. Additional types of RES comprise of waste heat derived from a renewable energy resource and used to produce electricity or useful, measurable thermal energy at a retail electric customer's facility or hydrogen derived from a renewable energy resource. Among the available RES, most of the power in the current generation mix are predominantly supplied from hydroelectric, wind, and solar resources. The rest is made up by other types of renewable energy sources.

11.1.1.1 Hydroelectric Power The generation of hydroelectric power is based on the conversion of kinetic energy from the falling water which drives the hydro turbines into mechanical power which is then converted into electric power by an electric generator. This technology is proven, mature, predictable, and cost-competitive. There are three main types of hydroelectric generation: run-of-river, storage, and pumped storage. The run-of-river plants have no storage capacity as their intake basins are small. In contrast, the storage hydroelectric power plants have a reservoir, and can be used for a broad range of energy services such as base load, peak load, and energy storage. In addition, these plants can be used as a regulator for other sources.

Pumped storage plants generate electricity mainly at peak demand periods. At off-peak periods when electricity prices are low, these plants pump the water at the low ground back into the reservoir at the high ground. Therefore, they act as both generator and load at different times of the day. Hydroelectric power plants have fast response which makes them suitable for providing regulation service. Currently they are considered as a type of grid energy storage with the largest capacity.

11.1.1.2 Wind Power After hydroelectric generation, wind power is considered to be the most cost-effective renewable energy technology. The wind power is a process by which the wind is used to generate mechanical power or electricity. Wind turbines convert the kinetic energy in the wind into mechanical power. This mechanical power is then converted into electricity through a generator to provide power to final consumers. Many individual wind turbines are clustered into a collection, called a *wind farm*, which is connected to the power transmission grid.

Wind speed is the single most important factor in determining the power output from a wind turbine. At very low wind speeds, there is an insufficient torque exerted by the wind on the turbine blades to make them rotate. However, as the speed increases, the wind turbine will begin to rotate and generate electrical power. The minimum speed at which the wind turbine first starts to rotate and generate power is called the *cut-in speed* and is typically between 3 and 4 m/s. As the speed increases above the rated output wind speed, the forces on the turbine structure continue to rise and, at some point, there is a risk of damage to the rotor. As a result, a braking system is employed to bring the rotor to a standstill. The speed at which this braking system kicks in is called the *cut-out speed* and is usually around 25 m/s. Wind turbines are susceptible to physical damage when faced with severe weather.

The wind power technology is already mature and well established. Onshore wind is an inexpensive source of electricity which is cost-competitive with coal-fired or gas-fired plants. Offshore wind is stronger than onshore wind and thus has higher potential for more wind power but the construction and maintenance costs of offshore wind are higher. However, the integration of wind power into the interconnected systems is not fully mature. Recently, the capacities of wind power resources have increased while their investment costs have declined. These new advances made possible for the significant reduction in the generation cost of wind power resources in the last few years.

11.1.1.3 Solar Power

Solar power is energy from the sun that is converted into electric energy. Solar power can be produced by using two types of technology: solar photovoltaic (PV) and concentrating solar power (CSP) technologies. The PV generators generate electricity directly from the sunlight using the electronic properties of certain type of materials which are embedded into the interconnected cells. The amount of energy that can be produced from a PV generator is directly dependent on the intensity of available sunshine (aka solar irradiance), the angle at which solar PV cells are oriented, and the ambient temperature. Initially, the investment cost of this type of technology was quite high. However, its average price has dropped considerably in the recent years, making the PV generation more cost-competitive.

CSP technology utilizes focused sunlight. CSP plants generate electric power by using mirrors to concentrate the sun's energy and convert it into high-temperature heat. That heat is then channeled through a tower which houses a boiler. The concentrated heat from the boiler generates steam which spins a turbine, which then drives a generator to produce electricity. Basically, CSP technology replaces the use of fossil fuels to heat the water. All CSP plants require large areas of land for collecting solar radiation to produce electricity at a commercial scale.

11.1.1.4 Other Renewable Energy Sources

Among the other types of smaller-scale renewable energy sources, it is worthwhile to mention two more types: biomass and geothermal.

As a renewable energy source, biomass energy can be produced by two different methods: (1) biomass energy that is produced by directly burning materials derived from vegetal plants and (2) biomass energy produced by burning gas released from the breakdown of organic matter. There are several types of biomass energy sources: food crops, agricultural waste, animal waste, wood waste, industrial wastes, any organic component, spent pulping liquors, combustible residues, combustible liquids, combustible gases, energy crops, or landfill methane. Although there are several techniques developed to produce energy from biomass, the most-widely used method is the direct combustion. The principle behind producing electricity from biomass energy is very similar to that of the thermal electricity generation. There is a lot of commercially proven power generation technologies that can use biomass materials as a fuel input. Biomass can be used as the unique fuel or can be combined with other sources of fuels for producing electricity from a combined cycle generator.

Geothermal resources consist of thermal energy from the Earth's interior stored in both rock and trapped steam or liquid water. As the thermal energy, the steam drives a turbine that rotates a generator which produces electricity. Electricity generated from geothermal energy is used for baseload power as well as peak load power. Available energy from biomass and geothermal sources is typically smaller than those from hydro, wind, and solar power sources.

11.1.2 Growth Drivers of Renewable Energy Sources

There are two main drivers which encourage the use of electricity generation from renewable sources. The two key drivers are the subsidization, and the carbon pricing. Nowadays, the subsidies to the renewable energy generators are the principal form of support to the growth of RES. There are two main forms of subsidies applied to renewable energy generation; the Feed-in Tariff and the renewable portfolio standards.

11.1.2.1 Feed-in Tariff Feed-in Tariff is a policy tool that was conceived to encourage more development and installations of renewable energy resources. It sets a long-term agreement to the producers of renewable energy resources. It takes as a reference the generation cost, and, in general, discriminate among different energy resources. Initially, this scheme was implemented in the United States in 1978, under a congressional law known as Public Utility Regulatory Policies Act (PURPA) whose primary motivation was to have better energy utilization and to promote the use of renewable resources. Since then, many countries adopt this type of subsidy regulation to support the more inclusion of renewable resources in the generation mix. It is a policy that ensures the access to the network grid, provides a mechanism to establish long-term contracts, and sets cost-based purchase prices. Different types of renewable resources are compensated for supplying energy to the system based on their fixed cost with allowance for reasonable profits.

11.1.2.2 Renewable Portfolio Standards

In the United States, the renewable portfolio standards (RPS) are public policy goals implemented by states to increase the utilization of renewable resources in producing electricity. US states have been active in adopting or increasing RPS, and 29 states now have them. Figure 10.2 from Chapter 10 shows the RPS policies in the US. These standards require utilities to source a specified percentage or amount of electricity from renewable energy sources. The requirement can apply only to investor-owned utilities (IOUs) but many states also impose this requirement to municipalities and electric cooperatives (munis and co-ops) as well although their requirements are equivalent or lower.

Renewable energy policies help drive the nation's $36 billion market for wind, solar, and other renewable energy sources. These policies can be integral to many state efforts to diversify their energy mix, promote economic development, and reduce emissions. Up until now, 29 states, Washington, D.C., and three territories have adopted an RPS, while eight states and one territory have set renewable energy goals. Iowa was the first state to establish an RPS and Hawaii has the most aggressive RPS requirement. In many states, standards are measured by percentages of kilowatt hours of retail electric sales. Iowa and Texas, however, require specific amounts of renewable energy capacity rather than percentages and Kansas requires a percentage of peak demand.

The main difference between RPS and the feed-in tariff methodology is that RPS does not guarantee the purchase of all the renewable energy offered. One of the goals of RPS is to allow more price competition among different types of renewable energy, and among specific RPS programs.

11.1.2.3 Carbon Pricing

To address the issue of climate change, emissions of harmful pollutants such as SO_x, NO_x, and CO_2 should be reduced and the investment into cleaner sources of energy must be encouraged. One approach to achieve these goals is via carbon pricing. The objective of carbon pricing is to capture what are known as the external costs of carbon emissions, that is, the costs that the public pays for in other ways, such as damage to crops, health care costs from heat waves and droughts or property damage from flooding and sea level rise, and link them to their sources through a price on carbon.

A price on carbon can help shift the burden for the environmental damage back to those who are responsible for it, and who can reduce it. Instead of dictating who should reduce emissions where and how, a carbon price gives an economic signal to the polluters to decide for themselves whether to discontinue their polluting activity and reduce emissions, or continue polluting and pay for it. In this way, the overall environmental goal is achieved in the most flexible and least-cost way to society. The carbon price also indirectly stimulates increased application of clean technologies and market innovations, fueling new, low-carbon drivers of economic growth.

There are two main types of carbon pricing: emissions trading systems (ETS) and carbon taxes.

An ETS—sometimes referred to as a cap-and-trade system—caps the total level of greenhouse gas emissions and allows those industries with low emissions to sell their extra allowances to other industries with high emissions. By creating supply and demand for emissions allowances, an ETS establishes a market price for greenhouse gas emissions. The cap helps ensure that the required emission reductions will take place to keep the emitters (in aggregate) within their pre-allocated carbon budget.

As another type of carbon pricing, a carbon tax directly sets a price on carbon by defining a tax rate on greenhouse gas emissions or—more commonly—on the carbon content of fossil fuels. It is different from an ETS in that the emission reduction outcome of a carbon tax is not pre-defined but the carbon price is. The main assumption behind the carbon tax is that the higher the carbon tax, the larger the reduction of carbon emissions.

The choice of which policy instrument to implement will depend on national and economic circumstances. There are also other indirect ways of more accurately pricing carbon, such as through fuel taxes, the removal of fossil fuel subsidies, and regulations that may incorporate a social cost of carbon. Greenhouse gas emissions can also be priced through payments for emission reductions. Private entities or sovereigns can purchase emission reductions to compensate for their own emissions (so-called offsets) or to support mitigation activities through results-based financing.

11.1.3 The Impact of Renewable Energy on Electricity Markets

In this section, the use of the term 'renewable energy' literally means wind and solar power sources only because wind and solar are predominantly the majority of fastest-growing RES in providing electricity to the grid. The availability of wind and solar power changes considerably through time because their power output depends on the varying weather conditions. In fact, the power fluctuation of these resources increases the variability of net system load. For example, wind power output depends on the current wind speed, and solar power output depends on solar irradiance. They also increase the net load uncertainty because the available power from wind and solar resources can only be partially predicted over the scheduled period. The difficulty associated with accurate prediction of their maximum available power affects the calculation of the market outcome and prices. Furthermore, the forecast uncertainties of wind, solar, and net load require the availability of additional resources so as to adjust their values in net load forecast calculations. This also affects the needs for additional ancillary services which may require the redesign of existing ancillary service markets.

Currently, the fixed capital costs of renewable energy technologies are relatively high. However, these resources have zero or near-zero variable production costs because of the almost-free source of energy. Moreover, in several market areas, renewable energy technologies are subsidized. Due to this subsidy, the variable costs of renewable energy resources can become negative. Negative variable cost means that the market will dispatch these resources first, reducing the energy production from other marginal resources, and as a consequence, reduce the energy prices. Clearly, the addition of supplies with zero or near-zero variable costs in the supply curve will shift the supply curve to the right thus causing the cost of electricity generation lower given the same demand. Due to the variability of RES output, the main task for renewable power producers is to adjust their contracts to match their power output at each period. The renewable energy suppliers can also set the market price during periods of low load when the renewable generation output is high, or during transmission-constrained periods, making the price of energy zero or negative. Since the generation from renewable resources is highly volatile, stochastic tools are needed for their scheduling process.

Since the renewable generators are installed in a geographically distributed wide area, full rated output from all plants is not concurrently achieved. This feature, in addition to the demand inelasticity, may increase volatility in energy prices. As the periods of high-energy output from renewable sources such as wind power may not coincide with the periods of high demand, these resources may not contribute directly to the requirement of power system resource adequacy. These resources are better suited as energy-only resources in contrast to reliable capacity resources such as conventional nuclear plants. Therefore, it is necessary to make some adjustments to the procedures used to determine the capacity contribution from such renewable energy sources to fulfill the resource adequacy requirement.

11.1.3.1 Capacity Values of Renewable Energy

Initially, several markets established regulatory standards to give capacity credits to the intermittent generation sources to compete with traditional generators. These credits are given based on the capacity of the generators. There are several alternatives for the capacity calculation. For example, due to the uncertainty of the generation produced by the renewable energy, some markets make a distinction between mature and immature generations. This is because generators with more than 3 years of operation provide a historical profile that makes the planning process easier. The capacity for mature generators is calculated based on the moving average for the past 3 years of the output power of the generation unit during critical periods; this period is set between 3 p.m. and 6 p.m. For the *immature* generators, a factor is applied to their capacity. This factor is calculated based on the measured capacity of each summer which is a known information. This factor corresponds to 20% of the rated power of the turbine.

The capacity factor for 1 year is calculated as the ratio of two quantities:

$$\mathrm{CF_T} = \frac{\displaystyle\sum_{t=1}^{T} P_{\mathrm{st}}}{\displaystyle\sum_{t=1}^{T} (P_{\mathrm{st}} + CN_t)}, \tag{11.1}$$

where P_{st} is the hourly power for each summer peak period t of the past 3 years, CN_t is the net capacity of each summer peak hour, and $\mathrm{CF_T}$ is the capacity factor for year T.

Then, the current capacity is calculated as the average of the capacity factor for the past 3 years. This calculation approach is based on the historical data with focus on the critical periods and with a moving average of the last 3 years. This method of calculation is done at the PJM market in the United States.

However, Mid-Continent Area Power Pool (MAPP) employs a different approach which is based on the monthly capacity values. In this method, the capacity is calculated based on the timing of its delivery during the peak load. If available, up to 10 years of historical data for both wind generation and load is used. Considering the monthly peak value, the values for the 4 hours around the peak period are selected. Then, the wind generation for these hours are sorted, and the median is calculated. This median value is set as the wind's capacity value for the month. The same calculation is performed for all years required and the final calculated value is used in operational planning in the power pool.

In ERCOT market, the peak period is considered as the duration between 4 p.m. and 6 p.m. from July to August during the summer. Therefore, the average output of renewable energy resources is computed based on the historical values during this period. In addition, a confidence factor to account for the variability of resource output is also considered. In all these approaches, the nominal capacities of wind and solar resources are discounted in a certain way while recognizing their actual contribution to meet the critical peak load. The approaches described above, consider the peak values for capacity calculations.

There are alternative methods in which the capacity calculation of renewable energies are based on the effective load carrying capability (ELCC). The ELCC is a percentile factor that describes the degree of performance of the generator source for meeting the reliability requirements, or any related problem such as an outage condition. This method can state if a resource, in this case a wind or solar generator, can serve the amount of an incremental load, considering two major components: the variability of the resource and the contingencies that produce a load not being served. The ELCC method was proposed by Garver in 1966. In his article, the author recommended the use of *loss of load probability* (LOLP) applied to ELCC calculation. However, there are growing interests in ELCC method due to the high volatility and high penetration of the renewable resources into the grid. The LOLP measure can be accumulated over a particular simulation time period, for example, a year or a month in which case this new factor is known as *loss of load expectation* (LOLE). Once the LOLE is determined, the ELCC of a particular resource can be determined. To account for the reliability effect of a particular renewable resource, the studied resource should be considered independently of the rest of resources in the ELCC calculation.

After the ELCC calculation, the renewable resource capacity is calculated as follows:

$$\mathrm{CF_T} = \mathrm{ELCC}\ (\%) \times P_{\mathrm{max}}(\mathrm{MW}) \tag{11.2}$$

All these statistical methods can be carried out using Monte Carlo Simulation methodology.

11.2 SMART GRID

Recently, the concept of smart grid (SG) has gained a lot of attention. On a high level, the goal of smart grid is to improve the delivery of electric power in a more reliable, efficient, responsible, and sustainable way. The main idea of SG is to migrate from a centralized structure to a more decentralized and consumer-oriented structure. If properly implemented, the SG can reduce the need for additional infrastructure such as transmission grid while keeping electricity supply reliable and affordable. From the electricity market point of view, SG provides options for adopting new technologies and motivating mechanism to produce alternative pricing incentives directly at the consumer level.

It is believed that the SG will allow the electric utilities to monitor the demand in a more detailed level, and to establish better demand response strategies. This dynamic control is possible in part by the application of smart meters (SM) connected at the point of consumption. The primary objective of this technology is to make the energy consumption more efficient so as to reduce the on-peak demand, and as a consequence, reduce the cost of energy. One of the expectations of the SG implementation is to make autonomous bilateral transactions in an electricity market but with regulations and operation by the system operator by providing an architectural platform to support end-to-end direct transactions between producers and consumers.

In this future scenario, consumers can have more options such as user's comfort and technological preferences, and thus an increasing participation by the demand response is expected. These options and developments will condition the consumer's response to the market price signals. However, the nature of the electrical loads limits the flexibility of the demand management for various reasons such as appliance duty cycles, the simultaneous applications of appliances, or user's predisposition to change routines. All these features representing electricity consumption patterns have been characterized by two control schemes based on the consumer price elasticity. The first control scheme represents the availability of the user to modify its energy use as a function of the electricity price and the second control scheme considers the technological restrictions which either allow or prevent demand pattern modifications by the consumers. The concept of SG is quite broad. It comprises of many other areas of developments which are further described below.

11.2.1 Microgrid

Microgrid can be any small autonomous power grid or smaller part of a larger power grid that can use its own generation to meet its own load within its operating area, independent of the larger grid. In the worst case scenario, the microgrid may rely on

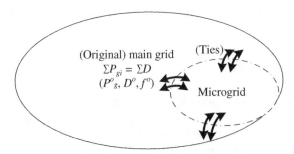

Figure 11.1 Microgrid in Grid-Connected Mode.

import generation from adjacent zones to meet its internal load. Microgrid must have the capability to balance its load and generation to maintain the desired voltage and frequency for reliability.

Conceptually, a microgrid can operate in either grid-connected or isolated mode. In the grid-connected mode, the microgrid is connected to the (typically larger) main grid. This can be the conventional way of system operation. The connectivity between the microgrid and the main grid can be strong, weak, or minimal. In the isolated mode, the operations of the microgrid and the main grid are independent of each other. This possible two-way *modus operandi* by a microgrid poses challenging, yet interesting questions in terms of market operation.

Figures 11.1 and 11.2 depict the two modes of operations with a microgrid connected or disconnected from the main grid, with the power balance equations (supply/demand balance) and desirable frequency.

Minimum characteristics of a microgrid should include, but not limited to,

1. Maintaining sufficient generation in its territory to meet its own load
2. Having necessary generator ramping and other capabilities, such as reactive power support, to maintain desired voltage and frequency
3. Establishing a market mechanism if pricing feature in the microgrid zone is desirable and necessary

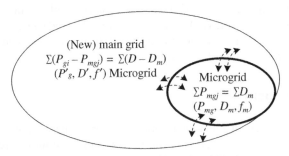

Figure 11.2 Microgrid in Isolated Mode.

As noted above, the key question to ask when a microgrid becomes part of the main grid, which has an operating electricity market, is in which mode the microgrid is operating and whether it will become a new market if isolated from the main grid. If the microgrid is still part of the main grid (grid-connected mode), the electricity market operation in this mode is the same as the previous, conventional market operation. However, if the microgrid operates in an isolated mode, a new market for the microgrid zone is desirable. In this case, the following conditions become necessary:

1. The original electricity market which encompasses the portion of original grid boundary but without the microgrid operating boundary has to continue to be operated without the microgrid as part of the market-clearing mechanism.

2. Network topology, generation, and demand have to be reconfigured or reupdated to reflect new grid structure (new main grid and microgrid).

3. Clear market rules have to be set up to determine which market participant, such as a physical generator, can participate in which market.

4. Two markets would be operated in either synchronous or asynchronous mode, treating the new market as 'contingency market,' as desired.

5. New market for newly formed microgrid boundary may need to be operated, with or without the connectivity with and the dependence on the remaining main grid.

There are several definitions and descriptions of microgrid in the literature. Readers can explore these models if they are interested. Different types of microgrids can appear in the near future.

11.2.2 Active Demand Response

As it is expensive to store electricity, the instantaneous power demand must be met with the instantaneous generation. As the demand varies with time, the market-clearing prices also vary. However, up until now, most of the final consumers connected at the distribution systems pay a fixed electricity rate which generally represents the average annual cost of service. Therefore, although the cost of electricity at the wholesale level changes over time, most electricity users pay a flat fee that does not reflect the wholesale prices at the time of consumption. Establishing mechanisms to efficiently manage the demand is not trivial. To address this problem, active demand response (ADR) should be supported in the future grid by providing the basic infrastructure that will allow the bidirectional flow of energy and communications among users, distributors, and marketers.

As defined in Chapter 10, demand response (DR) can be any resource that has the capability to change or reduce the electricity consumption at any given time. The mode to change the electricity consumption can be instantaneous or pre-scheduled. The key players of DR resources are those who consume, not supply, electricity. Typically, they are represented by residential, commercial, and industrial customers of electricity. If electricity consumption changes due to changes in the electricity prices as a function of time, this type of DR is called *price responsive demand* (PRD). Several research works have proposed different models and techniques applied to achieve demand management by consumers.

In order to motivate the electricity consumers to actively participate in the demand response market, several pricing schemes have been proposed. For example, some well-known DR pricing schemes include real-time pricing (RTP), day-ahead pricing (DAP), time-of-use (TOU), critical peak pricing (CPP), and inclining block rates (IBR). In all these pricing mechanisms, the main objective is to allow retail prices to reflect fluctuations in wholesale prices. Since electricity users will pay different prices at different times of the day, this will encourage them to shift the use of home appliances toward the hours that are outside of the peak load periods with the goal of reducing the electricity costs. This demand shift would also help reduce the ratio between the peak value and average value of electricity demand, known as peak-to-average ratio (PAR), so as to flatten the load curve. The main objectives of the variable pricing schemes are that users can reduce power consumption when the price is high and reduce their bills accordingly. Moreover, the peak demand can be reduced (*peak-shaving*), and the off-peak demand can be increased (*valley-filling*) which increases the capacity utilization of off-peak generation. The current tenet of these schemes is to allow active participation by electricity consumers with the goal of increasing price elasticity of demand.

Currently, time-varying pricing schemes are implemented in various regions of North America. For example, a DAP scheme hourly rate is used by Illinois Power, and TOU scheme with three levels of rates (on-peak, mid-peak, off-peak) is used by Ontario Hydro. The IBR scheme was widely adopted among the distribution companies since 1980s. For example, EPEC in Argentina, Southern California Edison, San Diego Gas & Electric, and British Columbia Hydro and Pacific Gas & Electric use a two-level tariff structure in which the marginal price in the second level is 40% or more higher than the first level, depending on the utility.

11.2.3 Impact of Active Demand Response

Currently, most DR studies focus on the analysis of system management either at the generation level or related to the market performance. However, with the paradigm change brought by the SG, it is clear that the primary benefit of ADR will depend on the feasibility studies for future DR deployments at the consumer domain. Therefore, it is important to discuss the DR issues in three different domains: generation, distribution, and consumers.

1. **Generation Domain**: Several studies of demand response are conducted from two different approaches. In the first approach, the DR management is incorporated endogenously considering elastic curves in market equilibrium models. The second approach considers the DR as an exogenous form that encompasses the variations in demand based on load reductions. More recent studies suggest stochastic methods that consider the randomness of the main variables involved.

2. **Distribution Domain**: FERC classifies the demand response based on how these programs affect the price of electricity and the corresponding time frame. Two main categories are defined, based on incentives and based on specified time. The incentive-based program is divided into classic and market-based programs. Within the classical program, two different approaches were further defined. The first approach is the interruptible demand program where participants receive incentives in advance or discounted rates. In this program, participants are asked to reduce their load values; otherwise, they will face penalties. The second approach is the direct control program, where some of the participating devices, such as air conditioners and water heaters are turned off remotely by the power company or the system operator.

   The other type of incentive program is defined as a market-based program. There are several sub-categories in this program: the supply of demand, the emergency services, and the ancillary service market. In the demand offering program, consumers specify the particular load that can be reduced. The bid is accepted if it is less than the market price; otherwise, the customer will not consume the load. In the emergency demand program, the consumers are paid compensation as a form of incentives to reduce the load during an emergency. In the ancillary service market, the load reduction is used to relieve congestion as well as to provide the auxiliary reserve. These schemes are available for some time now. However, they do not take advantage of what the SG can offer related to the two-way communication. This new feature improves the system optimization while keeping the comfort of home users. The new pricing schemes based on dynamic pricing scheme can become an alternative which effectively take advantage of this development.

3. **Customer Domain**: The study of consumer response to price signals is mostly confined in the areas of the technological comfort and user preferences. The evolution of the new pricing programs will be driven by the customer perception of the cost savings, and the sensitivity to issues related to the environmental impact. In addition, the evolution of the electric load characteristics will affect the demand response management. The effect can be represented

by two main variables that control the price elasticity of demand: first representing the availability of the customer to modify their consumption based on the fluctuating price, and the second related to the technological constraints that prevent or promote the change. The actual challenges are related to the demand volatility and inclusion of social factors and comfort sensitivity into the formulations.

11.2.4 Energy Storage Systems

With the increasing penetration of the variable type of renewable resources such as PV and wind energy, there are periods during the day when the RES output is greater than what is necessary to meet system demand. This additional energy in those surplus hours can be stored and used in other periods when there is a need for additional energy such as either a peak demand period or when the power is more expensive. This kind of analysis can be done at the utility side or customer side. If a customer can store excess energy during off-peak hours, he/she can sell this excess electricity back to the utility or another customer during peak hours. These ideas motivate the research and development in energy storage (ES) systems and their inclusion into the electrical systems. Therefore, ES devices will play a vital role within the SG framework. In addition to the peak-shaving and cost-saving issues explained above, there are additional important features associated with ES. If the storage devices are installed at the customer site, they can be used to provide voltage support or reduce congestion. Moreover, since certain types of ES devices such as electric vehicles can be deployed in different locations, these resources can be considered as mobile network compensator.

There is a wide variety of storage devices with many technical characteristics: energy capacity, power capability, charging and discharging cycles, efficiency, and time response. The most common types of ES are pumped hydro storage (PHS), compressed air energy storage (CAES), and flywheel energy storage (FES), which are used in bulk power systems. Batteries based on chemical or hydrogen can be considered in distribution systems and smart grid systems. Besides, double-layer capacitors (DLC) and superconducting magnetic energy storage (SMES) are additional types of storage devices.

11.2.5 Electric Vehicles

All electric vehicles (EVs) available in the market run at least partially on electricity. Unlike conventional vehicles that rely on gasoline or diesel-powered engine, electric cars and trucks use electric motors powered by electricity from batteries or a fuel cell. Not all electric vehicles work the same way. The plug-in hybrid electric vehicle (PHEV) offers both a gasoline or diesel engine and an electric motor which is powered by a battery that can be charged/recharged by plugging into a source of electric power.

Other EVs forgo liquid fuels entirely, operating exclusively on electricity. The plug-in electric vehicle (PEV) is an electric vehicle that runs entirely on battery-powered electricity. The depleted battery is recharged using electricity either from a wall socket or a dedicated charging unit. Since they do not run on gasoline or diesel and are powered entirely by electricity, PEV or battery electric cars are considered "all-electric" vehicles. Fuel cell electric vehicles (FCEV) convert hydrogen gas into electricity to power an electric motor and battery. Fuel cell vehicles are a relatively new technology in passenger vehicles, but have a substantial carbon reduction role to play alongside other all-electric vehicles. Generally, EVs can be integrated into home energy systems, as well as the electric grid. Massive adoption of these vehicles can promote the electrification of transportation system.

In addition to the regular use of electric vehicles as a transportation agent, EVs would use electricity from the grid, preferably during off-peak hours, nights and weekends, to charge, then discharge it back into the grid at other times. In the process, the vehicles could also provide regulation service to the grid, as needed. This concept is also known as vehicle-to-grid (V2G).

Electric vehicles (EVs) can also be considered as a storage device. Initially, they have been promoted to reduce the emissions produced by the vehicles based on internal combustion engines. Current EV technology uses high-performance batteries as the power source, such as nickel cadmium, nickel metal hydride, or lithium ion batteries. The EV batteries need to be charged by plugging them into the grid. Therefore, considered as a storage device, EV can be connected to the house and/or used as a combination with any other renewable resource such as PV as a generation resource or demand.

The challenge of incorporating these vehicles in market operation stems from a number of factors. First, from the wholesale market perspective, there must be a sufficient number of electric vehicles in the system to have an impact in terms of both load and generation. Second, in order to provide regulation service, these vehicles must be available at all times or at certain agreed-upon times with sufficient capacity. Third, we must make a distinction between capability and actual performance. The issue of capability versus performance is not new as can be evidenced from the same issue that is prevalent in existing generators, large and small, which participate in existing electricity markets.

The concept of sufficient number of vehicles is important. Since the charging/discharging capacity of each electric vehicle is typically small (in kW range), certain level of aggregation is evidently needed for meaningful participation in the wholesale market. For example, it would require about 800 to 1000 electric vehicles to achieve 1 MW of charging/discharging capacity, depending on the battery capacities of the vehicles. Since these vehicles are mobile, they also act as mobile load/mobile generation, which changes in both time/space dimension and magnitude. If the behavioral aspect of owners/drivers are considered, this poses a significant challenge to both market operation and simulation. The fundamental question would be that how the mobile load and mobile generation, represented by such vehicles be represented in day-ahead market and/or real-time market to accurately account for their unique physical and economic characteristics in market operation.

Participation of electric vehicles in an electricity market, based on either fixed-price or varying (dynamic) price, on either fixed-time or varying time for charging and discharging, could also pose some challenges to market operation. Forecast of their potential load/generation pattern would require more sophisticated forecasting tools and methods. This is similar to the challenges faced when forecasting the power generation from wind resources. The importance and immensity of this challenge is also proportional to the numbers and types of electric vehicles in the system of interest.

Up until now, the market for electric vehicles is not yet mature because the infrastructure is still inadequate to make this market attractive or profitable. However, this scenario can change in the future. Therefore, the current economic advantages of the EVs are in the area of saving due to their peak-shaving or valley-filling capability.

11.2.6 Smart Meters

To provide better flexibility for SG deployments, it is essential to have a good and reliable communication or interface system between the utility and the customers. The smart meters (SM) will play a crucial role in this regard. From the market perspective, having access to the real-time data related to electricity flows and prices is necessary for both the consumers and the utility. Without smart meters, customers do not have the information regarding their consumption in real-time, or the consumption information is only recorded monthly or even worse bi-monthly. Therefore, it is very difficult for customers to make intelligent decisions regarding energy costs or savings based on this late information which also lacks the details about their power consumption. That is one of the key reasons for the growing interest in the development and implementation of advanced smart meters for both utilities and consumers.

The desirable features of smart meters should include an ability to visualize the new architecture of tariff and pricing schemes. The installation of smart meters allows the transition from the manual to automatic energy consumption reading, fast remote communication, better consumption estimates, remote control of de-energization or re-energization process, or provision of better information for market process review. Smart meters installation is not new. In the United States, it started between 1980s and 1990s; however, it increases significantly in recent years due to the SG development. Several countries enacted legislations to regulate the installations of SM. The European Union envisions that SM will eventually replace the current meters and by 2020, almost 80% of the electricity meters will be SM. In other words, around 200 million SMs will be installed in Europe. Therefore, it is expected that smart meters will become a key vehicle in providing the vital information for electricity markets.

11.2.7 Electricity Market at Distribution System Level

Up until this point, various aspects of the electricity markets, such as design, operation, and pricing methods, have been presented and discussed in various chapters. The electricity markets that were referred to throughout the book are markets that operate at the *wholesale* level. The trading of electricity at the wholesale level spans a very large area and sometimes crosses many states or national borders. The primary part of the power grid that facilitates these wholesale transactions is the transmission system.

In contrast to the wholesale electricity markets, retail electricity markets that exist in many states in the United States or many other countries operate at the distribution system level and are confined to a smaller area. In the United States, retail markets are regulated at the state level. State regulatory authorities regulate a distribution utility's cost and rate of return for use, maintenance, and planning of the distribution system. Retail markets are the final link in the electricity supply chain. Retailers buy electricity in wholesale electricity markets, package it with delivery services and sell it to customers. These retail sales can range from the service for a large manufacturing facility to small businesses and to individual households. This is typically the main interface between electricity industry and customers such as households and small businesses. Therefore, the electricity supply for end-use customers is obtained through the open, competitive wholesale electricity market or from utility-owned rate-based generation or combination of the two.

In states where full retail competition, often known as *"retail choice"* is provided, customers may choose between their incumbent utility supplier and an array of competitive suppliers, as opposed to being a captive customer to a single provider. Competitive retail suppliers provide a variety of service plans that give consumers and businesses flexibility in their energy purchases. They may also offer options to hedge against price fluctuations, more choices for alternative energy resources, and newer energy efficiency projects, among others. These opportunities allow customers and businesses to choose the services that best meet their needs. In most states providing retail competition, those customers who do not choose a nonincumbent competitive supplier are serviced by their incumbent utility through a service called *provider of last resort* (POLR) or *standard offer service* (SOS). The POLR or SOS supplier will then have to secure the needed power on the wholesale market through a competitive bidding process.

Thus far, there has been limited incorporation of DERs, demand response (DR), energy storage (DESS), and energy efficiency (EE) into the distribution system planning efforts. This is because system planners are appropriately conservative, and inclined to consider only resources that are well known and can be relied upon to meet projected system needs. Some DER technologies have yet to fully mature, and the use of certain types of DER for system reliability purposes is still relatively new. However, these types of resources are poised to flourish in the near future for many reasons. Specifically, there will be a proliferation of various sized DERs in both transmission and distribution systems, an increasing number of local renewable resource installations at end-use points, and load growth through electrification of transportation and other end-uses. Some of these transformations, such as the deployment of distribution-level photovoltaic (PV) systems, represent relatively minor quantities of generation to date. Even though the continued penetration of small-scale PV systems may have little impact on the current grid, more recently large amounts of rooftop solar have begun to create operating and economic issues for distribution systems. If PV is highly concentrated in certain areas, it can have a significant impact on the local distribution grid. The same effect is true for plug-in electric vehicles (PEV), which make a small overall impact but present significant challenges if they become concentrated in small areas.

The growing challenges at the distribution system will also influence the transmission system operated in conjunction with wholesale electricity markets. However, the voltage levels of these wholesale markets lie at the edge of transmission or sub-transmission systems, whereas the operation of the low-voltage distribution system is the responsibility of local electric utilities or distribution operator. Therefore, it is a challenge for wholesale electricity markets in effectively dealing with these developments. By the same token, there is a growing need to bridge the gap between the system operation both at the transmission level and the distribution level while recognizing the impact of these developments on the system as a whole. As the growing number of DERs will be connected to the grid, the distribution system may need a new design to ensure that new services from DERs are valued and utilized effectively and efficiently. The new kind of *distribution system operator* (DSO) construct is needed to assume the responsibility of maintaining supply and demand balance at the distribution level, as well as linking the wholesale and retail market agents. This new grid design should also include a possible market mechanism at the distribution level in which available, feasible, and cost-effective DER solutions become part of any distribution system planning efforts. These market designs should create opportunities for DER resources with reliable track records to compete on a level playing field with more traditional resources. Planners must become more fully conversant with the capabilities, applications, and costs of DERs. Only in this way, progress can be made toward using new solutions for emerging problems in distribution system planning and operation.

11.2.8 Smart Grid Policies and Programs

To improve the energy efficiency in this new paradigm of energy generation and distribution, new market designs are necessary. First, new policies and programs are needed to set the rules, to encourage the energy savings, and the efficient use of the energy. That is the reason for the growth of several programs which attempt to implement different types of instruments such as subsidies, tax incentives, and trades. Most of these programs are aimed to encourage the utilization of renewable energy so as to reduce the use of fossil fuel sources. These programs can be classified into multinational, national, state, and city programs based on the covered region. Multinational programs focus on solving two main issues:

1. Provide access to reliable and sustainable electricity to all the people; and achieve the target of providing universal energy access in the shortest possible period. It is considered that more than one billion people on earth do not have access to electricity and another billion do not have access to a reliable electricity grid.

2. Reduce major environmental problems. There are seven environmental concerns related to electricity generation: pollution, global warming, natural resource depletion, water disposal, water pollution, acid rain, and ozone layer depletion.

In order to promote actions related to electricity markets, it is necessary to discuss programs related to establishing the main policies and regulations to coordinate resources and motivate market players. The first of these programs was established by IEA, and was named "1-Watt Initiative." The main objective of this program, proposed by Meier in 1998, was to produce regulations and policies to reduce the energy consumption of electrical and electronic appliances when they are in standby mode. The target was to reduce this power to less than a watt. The main objectives of this initiative were the following:

- Device manufacturers should reduce standby power to 1 watt or less
- Reduce consumption of all appliances by more than 50%
- Develop standard test procedures and definitions

This idea was promoted by the IEA by taking international actions to reduce the standby power waste from electrical equipment. The main features were to

- Measure the leaking electricity
- Evaluate initiatives at international, national, and regional levels
- Develop an action plan

Even though these programs were voluntary initially, several countries such as Australia and South Korea established and implemented them as mandatory. For example, the *e-Standby Program* and the *Energy Efficiency Labeling Program* became mandatory programs in these countries. In the United States, these programs started out as voluntary programs such as the *Energy Star* program that include provisions to measure and monitor networked standby modes in televisions and displays. In the European Union, a *Code of Conduct on Energy Efficiency of Broadband Equipment* was introduced, and regulations are currently being amended to include provisions to ensure that network-connected projects include power management features and that networked standby power consumption is reduced.

Nowadays, the development of *smart* home devices which allows connecting the home appliances to the Internet, is facing new challenges. Consequently, more and more homes are becoming highly connected to the Internet. Hence, energy consumption will increase. The main requirement is that these appliances need to be always connected to the Internet if they need to receive and send data, so they cannot be in either disconnected or standby mode. For example, the TV appliances can download movies, series of shows, etc., when these shows are being broadcasted and when consumers are not watching them. Refrigerators will be connected to transmit and order groceries. This new scenario will force that the 1-watt initiative may not apply because there will be no device that will be in standby mode. It is estimated that the electrical energy consumption related to Internet communication technology represents more than 5% of the total consumption. This rate is expected to triplicate in next 20 years. These new concepts and developments will force policy makers and regulators to start implementing new policies related to smart grid applications. Nowadays, several countries adopted laws, standards, and regulations related to these matters. Some regulations are mandatory and some are voluntary. In addition, several pilot projects have been implemented around the world to investigate the effectiveness of these rules and regulations.

Recent activities involve legislations and regulations related to advanced metering infrastructure (AMI), demand response, network metering, and distributed generation.

11.2.9 Market Implementation

The electricity market in the context of smart grids will largely depend on the demand response and energy efficiency programs. Market implementation approaches differ based on the particular objectives to achieve. Among these objectives, the following can be mentioned:

- Provide a price signal to the users to reflect the market response and the energy cost
- Reduce the energy price paid by the customers
- Increase market penetration at any customer level
- Influence energy user response and habits
- Increment user knowledge regarding different programs, energy services, and energy price schemes

The first objective related to implementing a market for SG is the consumer education. Consumers should be educated of the rewards of market programs so as to motivate them to actively participate in the markets. The educational programs take priority before any other market programs are implemented. There are different ways to implement consumer education programs such as: (1) inclusion of an explanation with the bills, (2) face-to-face explanation, (3) training, (4) advertisements, and (5) promotions.

After the education step, additional methods can be implemented based on the fact that any change on the price signal will influence the whole energy market. Some of these approaches include the time-of-use, off-peak pricing, seasonal, promotional, or some incentives such as rebates, credits, or cash grants. The main challenge is to encourage the customer participation by providing incentives for both the efficient production and efficient consumption of energy. It is also necessary to provide a transparent and effective information about services, products, and costs related to market implementation for SG.

11.3 FINAL THOUGHTS

No one can predict the future with any certainty. There are only possibilities. The future of power systems and that of electricity markets are no exception. While it is difficult to see the actual state of the future in 10 or 20 years with any clarity, we can see the emerging developments looming on the horizon. As mentioned in this chapter, some emerging developments include increasing amount of renewable energy sources, distributed energy resources, smart grid, microgrid to name just a few. A few power systems have been moving towards an ambitious goal of meeting the system load with 100% RES. While such a goal may be possible for a small system, it is rather a goal that will be extremely hard to achieve for a larger power system. Several studies and analyses have pointed out a possible RES saturation point for several systems, that is the maximum percentage of RES in the resource mix that a particular system can handle without the need to change its operating or planning protocols. In other words, the system has to modify its operating rules and procedures if additional RES beyond this saturation point are to be included.

We are also going to see some changes in the distribution system. A growing amount of dispersedly located energy resources will take a larger share of power supply to the load consumers. These resources will be located closer to the consumers who can also own such resources. Once passive load consumers can become active with an ability to feed power back into the grid with the resources in their backyards. Power flow patterns can change dramatically due to these changes. Innovative market designs at the distribution system will be required to accommodate such changes. Changes in the wholesale market designs are also expected due to changes in the distribution system.

Regardless of changes to the power systems and electricity markets, the future will be full of surprises and unforeseen events that will be interesting to see. One familiar economist has warned us once that markets are voluntary arrangements: they could disappear if not done properly.

FURTHER READING

1. Kaltschmitt M, Streicher W, Wiese A, editors. *Renewable Energy: Technology, Economics and Environment*. Berlin, Heidelberg: Springer-Verlag; 2007.

2. Morales JM, Conejo AJ, Madsen H, Pinson P, Zugno M. *Integrating Renewables in Electricity Markets: Operational Problems*. New York: Springer US; 2014.

3. Milligan M, Porter K. Determining the capacity value of wind: a survey of methods and implementation. In: Proceedings of WINDPOWER 2005, May 2005.

4. Ackermann T, editor. *Wind Power in Power Systems*, 2nd edition. UK: John Wiley & Sons; 2012.

5. Mendonca M, Jacobs D, Sovacool BK. *Powering the Green Economy: The Feed-in Tariff Handbook*. Routledge; 2009.

6. Kim SG, Hur SI, Chae YJ. Smart grid and its implications for electricity market design. *Journal of Electrical Engineering and Technology* 2010;5(1):1–7.

7. Kumar R, Ray PD, Reed C. Smart grid: an electricity market perspective. In: Proceedings of Innovative Smart Grid Technologies (ISGT), 2011 IEEE PES, January 2011.

8. Ozturk Y, Senthilkumar D, Kumar S, Lee G. An intelligent home energy management system to improve demand response. *IEEE Transactions on Smart Grid* 2013;4(2):694–701.

9. Li S, Zhang D, Roget AB, O'Neill Z. Integrating home energy simulation and dynamic electricity price for demand response study. *IEEE Transactions on Smart Grid* 2014;5(2):779–788.

10. Vivekananthan C, Mishra Y, Ledwich G, Li F. Demand response for residential appliances via customer reward scheme. *IEEE Transactions on Smart Grid* 2014;5(2):809–820.

11. Yoon JH, Baldick R, Novoselac A. Dynamic demand response controller based on real-time retail price for residential buildings. *IEEE Transactions on Smart Grid* 2014;5(1):121–129.

12. Roozbehani M, Dahleh MA, Mitter SK. Volatility of power grids under real-time pricing. *IEEE Transactions on Power Systems* 2012;27(4):1926–1940.

13. Qian LP, Zhang YJA, Huang J, Wu Y. Demand response management via real-time electricity price control in smart grids. *IEEE Journal on Selected Areas in Communications* 2013;31(7):1268–1280.

14. Weranga KSK, Kumarawadu S, Chandima DP. *Smart Metering Design and Applications*. Singapore: Springer; 2014.

15. Gil HA, Lin J. Wind power and electricity prices at the PJM market. *IEEE Transactions on Power Systems* 2013;28(4):3945–3953.

16. Hoeven MVD. Energy efficiency simply makes sense. *The Journal of the International Energy Agency, Visualising the "Hidden" Fuel of Energy Efficiency* 2013;4:2.

17. Garver L. Effective load carrying capability of generating units. *IEEE Transactions on Power Apparatus and Systems* 1966;PAS-85(8):910–919.

Index

Electricity Markets: Theories and Applications, First Edition. Jeremy Lin and Fernando H. Magnago.
© 2017 by The Institute of Electrical and Electronics Engineers, Inc. Published 2017 by John Wiley & Sons, Inc.

IEEE Press Series
on Power Engineering

Series Editor: M. E. El-Hawary, Dalhousie University, Halifax, Nova Scotia, Canada

The mission of IEEE Press Series on Power Engineering is to publish leading-edge books that cover the broad spectrum of current and forward-looking technologies in this fast-moving area. The series attracts highly acclaimed authors from industry/academia to provide accessible coverage of current and emerging topics in power engineering and allied fields. Our target audience includes the power engineering professional who is interested in enhancing their knowledge and perspective in their areas of interest.

Printed and bound by CPI Group (UK) Ltd, Croydon, CR0 4YY

16/04/2025

14658346-0002